SpringerBriefs in Public Health

For further volume:
http://www.springer.com/series/10138

Brian Castellani • Rajeev Rajaram • J. Galen
Buckwalter • Michael Ball • Frederic Hafferty

Place and Health as Complex Systems

A Case Study and Empirical Test

Brian Castellani
Kent State University
Ashtabula, Ohio
USA

Rajeev Rajaram
Kent State University
Ashtabula, Ohio
USA

J. Galen Buckwalter
Institute for Creative Technologies
University of Southern California
Playa Vista, California
USA

Michael Ball
Kent State University
Ashtabula, Ohio
USA

Frederic Hafferty
Program in Professionalism and Ethics
Mayo Clinic
Rochester, Minnesota
USA

ISSN 2192-3698   ISSN 2192-3701 (electronic)
SpringerBriefs in Public Health
ISBN 978-3-319-09733-6   ISBN 978-3-319-09734-3 (eBook)
DOI 10.1007/978-3-319-09734-3

Library of Congress Control Number: 2014960229

Springer Cham Heidelberg New York Dordrecht London

Printed on acid-free paper

Springer is part of Springer Science+Business Media (www.springer.com)

About the Authors

Brian Castellani, PhD is professor of sociology and head of the Complexity in Health Group, *Kent State University*. He is also adjunct professor of psychiatry, *Northeastern Ohio Medical University* and advisory board member, Center for Complex Systems Studies, *Kalamazoo College*. Dr. Castellani is internationally recognized for his expertise in computational and complexity science method and their application to several key topics in health and health care, including community and public health. He is specifically recognized for his development (with Dr. Rajaram) of a new approach to modeling, called case-based complexity. And, he is known for developing the SACS Toolkit: a case-based, mixed-methods platform for modeling the temporal and spatial dynamics of complex systems, particularly in large and big data. For more on Castellani, see (http://www.ashtabula.kent.edu/about/facultystaff/faculty/~bcastel3/).

Rajeev Rajaram, PhD is associate professor of mathematics and research member of the Complexity in Health Group, *Kent State University*. As a mathematician, Dr. Rajaram is internationally recognized for his applications of stability and control theory of differential equations to modeling a variety of topics, including community and public health. He is also recognized, with Castellani, for his development of a case-based complexity approach and the SACS Toolkit. For more on Rajaram, see (http://www.kent.edu/math/profile/rajeev-rajaram).

J. Galen Buckwalter, Ph.D. is research scientist, Institute of Creative Technologies, *University of Southern California* and founding scientist, TidePool (http://www.tidepool.co/). Dr. Buckwalter has had an extremely active career both as an academic research scientist and as an entrepreneur in the private sector. His academic career, with over 100 peer reviewed publications, has focused on advancing the field of psychometrics and personality testing/assessment, and, more recently, resilience and allostatic load.

Michael Ball, MS is research coordinator, Complexity in Health Group, *Kent State University*. Mr. Ball is gaining an international reputation for multi-agent modeling and machine intelligence, primarily in application to disease transmission and community and public health. For more on Ball, visit (http://www.personal.kent.edu/~mdball/).

Frederic Hafferty, PhD is professor of medical education and professionalism, Mayo Clinic. Dr. Hafferty is world-renown for his work in medical professionalism, the sociology of professions and medical education. He is specifically known for his ground-breaking work in the hidden curriculum and its application to medical education, as well as his development (with Castellani) of a complex systems view of medical professionalism. For more on Hafferty, see (http://www.researchgate.net/profile/Frederic_Hafferty).

Preface

Over the last decade, scholars have developed a *complexities of place* (COP) approach to the study of place and health. According to COP, the problem with conventional public health research is that it lacks effective theories and methods to model the complexities of neighborhoods, communities and so forth, given that places exhibit nine essential "complex system" characteristics: they are (1) causally complex, (2) self-organizing and emergent, (3) nodes within a larger network, (4) socio-spatially dynamic and evolving, (5) nonlinear, (6) historical, (7) socio-spatially open-ended with fuzzy boundaries, (8) critically conflicted and negotiated, and (9) agent-based.

However, while promising, the COP approach is currently faced with two challenges: its comprehensive definition of complexity remains systematically untested; and its recommended computational and complexity science methods (e.g., geospatial modeling, social network analysis, agent-based modeling) have yet to be organized into a cohesive framework.

The current study therefore conducted an exhaustive test of all nine COP characteristics and suggested methods. To conduct our test we made two important advances: First, we developed and applied the *Definitional Test of Complex Systems* (DTCS) to a case study on community health and sprawl (a complex systems problem) to examine, in litmus test fashion, the empirical validity of the COP's 9-characteristic definition. Second, we employed the SACS Toolkit, which we used to organize the suggested list of COP methods into a cohesive framework. The SACS Toolkit is a case-based, mixed-methods platform that draws on the latest advances in computational and complexity science methods to model the temporal and spatial complexities of complex systems. For our case study we examined a network of 20 communities, located in Summit County, Ohio USA. In particular, we examined the negative impact that suburban sprawl is having on the poorer communities in this county. Our database was partitioned from the *Summit 2010: Quality of Life Project*.

Overall, we found the COP's 9-characteristic definition to be empirically valid and useful. We also found the SACS Toolkit to be an effective way to employ and organize the methods recommended by the COP approach. Nonetheless, some issues did emerge. For example, the COP approach seems almost entirely focused on the complexities of place. As such, it has yet to develop a sophisticated view of how place, health and health care are intersecting complex systems. Also, while it is useful to think of places as agent-based based (Characteristic 9), there are limits to this modeling approach, such as its microscopic view of emergent social structure and its restricted (rule-based) view of agency. Still, despite these challenges, the COP approach seems to hold real empirical promise as a useful way to address many of the challenges that conventional public health research seems unable to solve; in particular, modeling the complex evolution and dynamics of places, and addressing the causal interplay between compositional and contextual factors and their impact on community-level health outcomes.

Acknowledgements

We would like to thank Dean Susan Stocker and Kent State University at Ashtabula for all of their financial and administrative support.

Contents

Chapter 1
The Complexities of Place Approach

1.1 Background: From Composition to Complexity

Over the last decade, the community health science literature has gone through a series of shifts, moving from the study of composition to the *complexities of place*. This shift began in the early 1990s, when a growing network of researchers moved away from the traditional study of compositional variables to the independent contribution that socioeconomic context has on health [21, 23, 24, 47, 60]. In this new research, **compositional factors** are defined generally as the aggregation of individual-level variables, such as household income, age, ethnicity, educational level or occupation [53]. **Contextual factors** are defined generally as the geographical, cultural or socioeconomic conditions/opportunities in which people live [47]. Such factors range from air quality to job growth to a community's health care system [53]. In this new research, compositional factors are seen as relatively independent of (orthogonal to) context. Therefore, when measuring the impact of context, these researchers often partial out (control for) individual or household circumstances [47, 53].

Between the 1990s and the early 2000s, however, another shift took place, as the study of context conceptually widened into the study of place—the umbrella term for research exploring the role that socio-geographical-contextual forces have on health and wellbeing [21]. In these studies, the places explored included communities, counties, cities, neighborhoods, the built environment, social networks and poverty traps, as well as state-level and national-level differences in health outcomes [21, 62, 64].

The shift from the study of context to place reflected an increasing recognition that the initial conceptualization of context, including its distinction from composition, was theoretically and methodologically problematic [6, 21]. Macyntire and Ellaway [47] summarized this theoretical problem as such: "'Composition' and 'context' are frequently treated as unproblematic and obvious distinctions, and underlying causal models are often implicit" (p. 129). With little to no theoretical rationale, researchers have overemphasized the differences between context and composition, spending almost no time (causally speaking) thinking about how people and contexts go together to form places. In terms of method, the problem is that places and their health are too complex for linear modeling, variable-based analysis, and overly simplistic causal models that treat compositional and contextual factors as mutually

© Springer International Publishing Switzerland 2015
B. Castellani et al., *Place and Health as Complex Systems*,
SpringerBriefs in Public Health, DOI 10.1007/978-3-319-09734-3_1

exclusive categories. For researchers such as Cummins et al. [21] and others [47, 62], the failure to recognize the limitation of conventional methods explains the lackluster results that studies of context and health regularly achieve. Often researchers find that, after the effects of various compositional variables are partialled out of an equation, contextual factors have little to no effect [21, 53]. According to Cummins et al. [21], these weak results, in turn, lead researchers to employ increasingly sophisticated conventional methodologies, but without yielding much improvement in the results, including rather weak partial-correlation coefficients and "relative risk" ratings of less than 2.0 (See [53]).

1.2   Places as Complex Systems

The theoretical and empirical potential of contextual (place) research, combined with the failure of conventional theories and methods to procure this potential, led Dunn and Cummins to guest-edit a special edition of *Social Science and Medicine* (Volume 65, 2007). The purpose of the various articles in this special edition (which ranged from position papers to exploratory empirical studies) was to point toward new theories and methods that do a better job of "placing health research in context" in order to "address the unanswered questions about the importance of social and geographical context for health variation" [27]. The new theories embraced included a relational approach to place and health (which comes out of geography, structuration theory, informal reciprocity and actor network theory) and complexity science. The new methods included many of the latest developments in computational and complexity science method, including geospatial modeling, qualitative complexity, social network analysis, agent-based modeling, topographical neural nets, etc. In terms of unanswered questions, the issues all revolved around arriving at a rigorously defined theory of place that can (a) handle the complex (i.e., nonlinear, emergent, self-organizing, multi-level) causal pathways typical of places and (b) explain how these complex pathways impact health. For example, in terms of our reading of Dunn and Cummins' special edition [27], while arguments vary, a definitional theme exists, which can be stated as such: people and places need to be integrated; related, places need to be thought of in holistic or systems terms as complex, emergent entities; furthermore, places need to be seen as functioning at multiple levels of scale; operating with open-ended boundaries; fluid, mobile and evolving; not constrained by traditional notions of space and time; comprised of nonlinear feedback loops and causal pathways; with subjective histories and multiple social meanings; emerging out of the intersection of the micro and macro, the local and global, and agency and structure; and, finally, as nodes in a larger network of places and environmental forces. In short, places need to be treated as complex systems.

Around the same time that Dunn and Cummins published their special edition, a handful of related articles and books (a few of which were published by the authors included in Dunn and Cummins' special edition) made a similar call. The main difference in these publications, however, was that complexity science, both theoretically and methodologically, was front and center in their argument [2, 7, 24, 25, 30, 34, 40]. They also explicitly theorized place (a.k.a. context, schools, communities, cities, counties, countries) as a complex system. Let us explain.

The cornerstone concept of complexity science is the complex system. As Cilliers [19] and others demonstrate (e.g., [5, 14, 16, 41, 49]), the definition of a complex system is not dictionary in form; instead, it is encyclopedic, and for good reasons. Almost every review of complexity science begins by listing the characteristics central to "their" definition of a complex system. Complex systems, it is variously argued, are self-organizing, adaptive, emergent, comprised of a large number of elements, autopoietic, nonlinear, dynamic, network-like in structure, open-ended with fuzzy boundaries, interdependent, agent-based, evolving, chaotic, comprised of feedback loops, historical, nodes within a larger network of systems, environmentally impacted, etc ([11, 16, 19, 49]). These encyclopedic listings are crucial because, in terms of the theoretical rigor of complexity science, the explanatory power of a complex system's definition (qua theory) comes from its characteristics [19]. Together and in isolation, the characteristics of a complex system not only describe what it is; they explain how it works.[5, 14, 16, 41, 49]

In terms of community health science, for example, to say that a place is emergent is to describe and explain how that place works—or at least that is the argument complexity scientists make [49]. Each characteristic listed in a definition constitutes therefore a whole domain of inquiry. The concept of emergence, for example, connects to a line of research that extends back to the late 1800s and classical sociology, systems biology and Gestalt psychology [16], as well as, more recently, to systems science and cybernetics [35], and, starting in the 1980s, to the beginnings of complexity science as defined by the Santa Fe Institute [45, 68] and its key scholars such as Holland [36] and Kauffman [39].

In terms of theorizing places as complex systems, the scholars doing this work have offered their own encyclopedic definitions. Gatrell [30], for example, offers the following definition of place—which he borrows heavily from Cilliers [20]. As a complex system, places are comprised of a large number of elements, interacting dynamically across networks, with rich, short-range interactions that can have a widespread impact; where each element (agent) is ignorant of the behavior of the place as a whole; were interactions are generally nonlinear (e.g., feedback loops, etc); and where place is emergent, open-ended, self-organizing, operating in a position far from equilibrium and evolving.

Despite the increased emphasis that Gatrell [30] and colleagues put on complexity science, the "rigorously defined theory of place" they articulate is essentially the same as the relational approach. This similarity is due, in large measure, to the latter's usage of complexity science as well, albeit more informally. For example, both Gatrell and the relational approach emphasize the agent-based, nonlinear, open-ended, emergent, "network like" structure of places. In fact, if Dunn and Cummins' special edition is combined with these related articles, an overarching theoretical definition of "place as a complex system" emerges.

Table 1.1 Communities as complex systems: list of key characteristics

As a complex system, communities are

1. Causally complex (i.e., circular causality, feedback loops, concurrent events taking place at multiple levels), such that context and composition are interdependent
2. Case-based, complex configurations that self-organize and emerge out of the compositional, contextual and health outcomes factors of which they are comprised
3. Nodes within a larger network of communities and social forces, such that more attention should be given to the socio-spatial position of places relative to each other and to macroscopic social forces
4. Dynamic and evolving, usually along various socio-spatial trajectories and around different attracting clusters. Given this perspective, use more dynamic terms such as "declining" or "getting better."
5. Nonlinear (e.g., poverty traps) both in dynamics and in the impact health interventions, in the form of public health policy have
6. Historical (e.g., institutional memory) and phenomenological (i.e., people have their own subjective, interpretive frames for understanding and participating in their communities)
7. Spatially and sociologically open-ended with fuzzy boundaries, such that communities are nodes within larger flows of people, places and things
8. Comprised of conflicted, negotiated power struggles amongst its major players and key subsystems, such as its community leaders or health care systems, etc.
9. Agent-based; comprised of a large number of interacting agents; with agents being mobile and evolving. And yet, while agent-based, complex systems are emergent and selforganizing

As shown in Table 1.1, this overarching definition (according to our reading of the literature) includes a total of nine major characteristics. Note: From here on out, we will refer to these characteristics, in their entirety, as the COP's 9-characteristic definition; also, we will refer to each characteristic in abbreviated form – for example, Characteristic 9 (the idea that places are agent-based) will be labeled C9.

With this note made we continue: According to the COP's 9-characteristic definition, places are defined as (1) case-based, causally complex configurations (e.g., multi- dimensional, multi-level, feedback loops) that (2) self-organize and emerge out of the compositional and contextual factors of which they are comprised; furthermore, (3) they are nodes within a larger network, such that their social and spatial position and place near one another is important in terms of health and wellbeing; (4) they are dynamic and evolving across time/space; (5) they are nonlinear, particularly in terms of how health policy interventions impact them; (6) they are historical and phenomenological; (7) they are spatially and sociologically open-ended, such that they are nodes within a larger flow of people, places and things; (8) they comprised of conflicted and negotiated power struggles amongst their key subsystems and major players; and (9) they are agent-based, emerging from the ground-up.

An overarching methodological theme emerges as well, which is as follows: variable-based, linear statistics cannot adequately model the complexities of place, leading instead to poorly designed studies (e.g., overly developed hierarchical regressions), false conceptual distinctions (e.g., treating compositional and contextual variables as orthogonal) and incorrect results (e.g., context is marginally important). To correct this problem, new methods, such as those found in complexity science (e.g., computational modeling, complex network analysis, agent-based modeling, geospatial modeling), are needed. Equally important, a new methodological framework is necessary, grounded in an epistemology of complexity and complex systems ([2, 62]).

1.3 The Research Problem

There is, however, a problem with the COP approach and its critique of conventional theory and method. Its COP's 9-characteristic definition has yet to be rigorously defined and empirically tested; and its methodological advances have yet to be extensively used or integrated into a cohesive methodological approach.

It has been roughly seven years since the 2007 Special Edition of *Social Science and Medicine* was published. During that time the *complexities of place* approach has somewhat taken off. Google Scholar shows over 410 citations to the 2007 special edition article by Cummins et al. A similar search using the terms "complexity theory" and "public health" gets roughly 850 hits; and a search using the terms "complexity theory" and "community health" gets roughly 190 hits. Also, in 2013, *Social Science and Medicine* published another special edition (volume 93); this time devoted to the broader topic of complexity theory and its capacity to advance the study of health and health care systems. The articles in this special edition cover a variety of topics, from the complexities of health care interventions to the 'systems' nature of various epidemiological phenomena to the complex intersecting relations amongst places, health and health care. Equally important, Sturmberg and Martin published (with Springer) the first, edited compendium of the work taking place at the intersection of complexity theory and health and health care research. The resulting 954 page tome is called the *Handbook of Systems and Complexity in Health* [66].

While neither of these 2013 publications are devoted explicitly or specifically to the *complexities of place*, they do demonstrate that the complexity sciences and, more specifically, complexity theory are making major inroads into the study of health and health care. Still, despite these significant inroads, the theoretical assumptions underlying a *complexities of place* (COP) approach remain untested in the social science and medicine literature; and the methods the COP approach employs (particularly its computational and complexity science techniques) remain underused or integrated into a cohesive methodological framework. Nonetheless, to be fair, some work has been done on these two challenges([21, 24, 25, 30, 34, 40]). Still, despite this work, as of 2014, the majority of scholarship has focused on a very specific topic or has been positional/speculative in nature, drawing its theoretical and empirical support from disparate areas of inquiry done on this or that aspect of the COP definition, or from fields outside community health. This lack of comprehensive testing is problematic because, without hard won, nose-to-the-ground, critical data analysis of the COP approach as a whole, it is difficult to explain exactly how studying places as complex systems is an improvement over conventional research. Furthermore, as the 2013 special edition of *Social Science and Medicine* makes clear, researchers uninterested in or skeptical of these ideas can easily dismiss them – which, in many ways, means the perpetuation of conventional research and health policy, despite its significant limitations.

1.4   Research Solution

The purpose of the current article is to advance the public and community health science literature by conducting an exhaustive test of the empirical validity and theoretical utility of the COP approach, as pertains to a case study on *sprawl and community-level health*. The database used for the current study was partitioned from the *Summit 2010: Quality of Life Project* (Summit-QLP) [67]—see Methods for more information. Using this database, the current article conducted an exhaustive diagnostic test of all nine characteristics and suggested complexity methods. It also made two important advances: it introduced a new diagnostic test for testing the utility of complexity science definitions and a new complexity-science method, called the SACS Toolkit, for advancing a cohesive, mixed-methods platform for modeling the complexities of place. Let us explain.

1.4.1   Sprawl in the States: A Complex Systems Problem

For the past few years we have been examining the negative impact sprawl has had on the health and wellbeing of a network of 20 communities in Summit County, Ohio—a typical Midwestern county in the United States that has been hit hard by post-industrialization and the outmigration and clustering of affluent people and resources from the cities into the suburbs.

For us, sprawl in the States is a good case study to test the COP approach because it is, by definition, a complex systems problem. Let us explain.

Sprawl in the States is not something that happens to one community or town. Instead, it is a 'complex systems' phenomenon spread out across, and emerging out of the networked evolution of multiple places across time and space.

Equally important, sprawl in the States emerges out of the complex interplay between compositional and contextual factors. In terms of composition, sprawl is a series of microscopic behaviors engaged in by a network of individual agents: families, businesses, etc. More concretely, at the micro-level, it is the semi/unplanned out-migration (flow) of relatively low-density development and residential relocation from urban centers into the suburban and semi-rural tiers surrounding an urban area. The relatively unplanned nature of sprawl in the States comes from the fact that, like many complex systems, no single force or agent is steering it. Instead, the system is evolving, self-organizing and emerging on its own, the result of a large number of adaptive, self-focused agents, across different communities, interested in upward social mobility. In turn, from the top-down, sprawl also involves the various macroscopic, community-level arrangements in which the above micro-level behaviors take place. These arrangements constitute the various social institutions operating within and across communities, from health care systems to regional economies. Community health science refers to these macroscopic behaviors and structures as contextual factors.

As sprawl in the States evolves, it creates a geographical network of segregation and exclusion where communities become, in many ways, relative islands in terms of resource usage, politics, wellbeing, etc, with movement amongst communities being largely automobile dependent. A significant, macroscopic consequence of this segregation is that the poorer, urban communities in a place like Summit County, which already have unsatisfactory health outcomes, do not improve ([22], [60]); or worse, they fall into what Bowels, Durlaff and Hoff [8] call a socio-spatial poverty trap: a self-reinforcing situation of persistent and intractable poverty.

When we put all of the above together, we arrive at the following definition: in the States, sprawl is a self-organizing, largely unsupervised, nonlinear, dynamic, negotiated ordering of multiple places, all networked together, evolving across time, which emerges out of (a) micro-level, agent-based out-migration and low-level development and (b) the interplay of these micro-level compositional forces with the various and different macro-level structural arrangements (contextual forces) existing across, within and between the evolving network of communities being studied. In short, sprawl in the States is a complex systems problem.

The question, however, in terms of the current study, is as follows: is the COP literature better than convention at defining and modeling this complex systems problem? Our argument was that if, in the process of testing the COP approach, this approach arrived at new and novel information about sprawl in the States, then this approach would be deemed theoretically and methodologically valid.

1.4.2   Overview of Study

The design of our study is a variation on a study by Keshavarz et al. [40]. To our knowledge, this is the only test-case in the public health literature that has sought to empirically validate, in an exhaustive manner, the usage of a complex systems framework. In their study, Keshavarz et al. [40] used a mixed-methods design (qualitative method and document analysis) to "examine the relevance and usefulness of the concept of 'complex adaptive systems' as a framework to better understand ways in which health promoting school interventions could be introduced and sustained" ([40], p. 1467). For their study they combed through the general literature on complexity science to arrive at a working definition of schools as complex systems. Next, they chose their case study, a set of public schools that had "implemented at least one health promoting schools project." Then, to test the empirical validity of their definition they explored, in litmus test fashion, each of its key characteristics to determine if their case study did, indeed, exhibit the characteristics of a complex system.

Like Keshavarz et al. [40], as explained above, our study conducts a litmus test of a case study: sprawl and its impact on the macroscopic health outcomes of a network of 20 communities located in Summit County, Ohio USA. However, our study differs in two important ways:

First, our intention is different. Keshavarz et al. [40] wanted to see if the data they collected empirically evidenced the characteristics outlined in their definition. From this perspective, they assumed their definition and its characteristics to be reasonably valid. For them, the issue of validity had to do with whether the case study fit their definition. The intention of our study is the opposite: we wanted to see if the characteristics in Table 1.1, as defined, should be applied to our data, to determine their degree of fit. In other words, our question was: Does the definition fit the case study? For us, this difference in intent highlights the major problem in complexity science research today, as applied to the social sciences: the definitions of complex systems that social scientists use are generally assumed valid; the only question to be answered is "does the data fit?" However, based on the extensive critiques made by Cilliers [19] and others (e.g., [12, 13, 49]), in the social sciences the real challenge is the opposite: determining the validity of the definitions used, as applied to each and every topic of study. Testing definitions is not, however, a matter of epistemology. The critiques listed above acknowledge the theoretical and methodological plurality of complexity science. Instead, testing is a matter of fit. Is the chosen definition (metaphorical or not) empirically valid and theoretically valuable?

Given our difference in intent, for this study we developed and employed the *Definitional Test of Complex Systems* (hereafter referred to as the DTCS). The DTCS is a formal test that guides researchers through the process of assessing the empirical validity of defining their topic as a complex system—see method section for more information.

Second, our methodology is different. In a follow-up article, Haggis engaged in a methodological critique of Keshavarz et al. [40] Haggis's [34] critique echoes the second challenge that complexity science makes to the conventions of community health science: if places are complex systems, then new methods and, more important, methodological frameworks are necessary. As pointed out in our introduction, the critique is as follows: qualitative analysis or statistics alone cannot get the work done, as qualitative analysis cannot handle the large, complex, multi-dimensional, multi-level databases typical of complex systems; and statistics, alone, cannot adequately model qualitative variables, nonlinearity or systems-level causality. Equally important, if one is to employ the new methods of computational and complexity science modeling, some sort of cohesive methodological framework is necessary.

Given this methodological challenge, for this study we advanced a case-based complexity science approach to studying places and their health. Specifically, we used the SACS Toolkit to conduct our test. The SACS Toolkit is a case-based, computationally grounded, theoretically driven, cohesive, mixed-methods toolkit for modeling complex systems [17].

Based on our two advancements, our study is organized as follows. We begin with method, where we review our dataset, measures, the DTCS and the SACS Toolkit and the specific techniques used for the current study. Next, we turn to our results, where we employ the DTCS and the SACS Toolkit to examine, in litmus test fashion, each of the nine characteristics listed in Table 1.1, as applied to our case study on sprawl in the States, to determine the empirical validity and theoretical utility of defining place and health as a complex system. Our goal will be to see if and how the COP approach (both its definition and methods) results in new and novel information about sprawl and its impact on the evolving macroscopic health outcomes of a network of communities. If it does, we can assert with some degree of confidence that the COP approach has a degree of theoretical and methodological validity. Finally, we end by summarizing the results of our test and suggesting how the DTCS and SACS Toolkit can be used by other researchers.

Chapter 2
Definitional Test of Complex Systems

Created for the current study, the Definitional Test of Complex Systems (DTCS) is our attempt at an exhaustive tool for determining the extent to which a complex system's definition fits a topic. The DTCS is not, however, a standardized instrument. As such, we have not normed or validated it. Instead, it is a conceptual tool meant to move scholars toward empirically-driven, synthetic definitions of complex systems. To do so, the DTCS walks scholars through a nine-question, four-step process of review, method, analysis, and results—see Table 2.1.

Following Gatrell, the DTCS does not seek to determine if a particular case fits a definition; instead, it seeks to determine if a definition fits a particular case. As Gatrell explains [30], the challenge in the current literature is not whether places are complex systems; as it would be hard to prove them otherwise. Instead, the question is: how do we define the complexity of a place? And, does such a definition yield new insights? Given this focus, Question 9 of the DTCS functions as its negative test, focusing on three related issues: the degree to which a definition (a) is being forced or incorrectly used; (b) is not a real empirical improvement over conventional theory or method; or (c) leads to incorrect results or to ideas already known by another name. Scholars can modify or further validate the DTCS to examine its further utility. Let us briefly review the steps of the DTCS:

Step 1: To answer the DTCS's initial five questions, researchers must comb through their topic's literature to determine if and how it has been theorized as a complex system. If such a literature does exist, the goal is to organize the chosen definition of a complex system into its set of key characteristics: self-organizing, path dependent, nonlinear, agent-based, etc. For example, if our review of the community health science literature, we identified nine characteristics. If no such literature exists, or if the researchers choose to examine a different definition, they must explain how and why they chose their particular definition and its set of characteristics, including addressing epistemological issues related to translating or transporting the definition from one field to another.

Step 2: Next, to answer the DTCS's sixth question, researchers must decide how they will define and measure a definition and its key characteristics. For example, does the literature conceptualize nonlinearity in metaphorical or literal terms? And, if measured literally, how will nonlinearity be operationalized? Once these decisions

© Springer International Publishing Switzerland 2015   11
B. Castellani et al., *Place and Health as Complex Systems*,
SpringerBriefs in Public Health, DOI 10.1007/978-3-319-09734-3_2

Table 2.1 Definitional test of complex systems (DTCS)

Step 1: Literature review and formulation of the definition

Question Set 1: What definition of a complex social system will be used?
1. What is the definition?
   a. For example, is the definition dictionary in form or encyclopedic?
   b. What are its key characteristics?
2. Where does the definition come from?
   a. For example, is the definition currently used in the field, or is it a new definition?
3. What are the definition's epistemological assumptions?
   a. For example, is it postmodern, critical realist, naïve realist, constructionist, etc?
4. What is the theoretical basis for the definition?
   a. For example, is the definition meant to be metaphorical, literal or prescriptive or some combination?
5. Does the definition or any of its key characteristics seem to be empirically or theoretically problematic?
   a. For example, are there examples in the literature where usage of the definition led to (i) poorly designed studies, (ii) faulty empirical results, or (iii) flawed or unclear theoretical conclusions?

Step 2: Methods

Question Set 2: How will the definition be operationalized and tested?
6. How will the current test be conducted?
   a. For example, what measures will be used?
   b. What case study will be used?
   c. What analytic techniques will be used for the test?

Step 3: Run test

Step 4: Determine results

Question Set 3: What conclusions about the validity and value of the definition were determined?
7. Did the test suggest that the definition is empirically valid?
8. Did the test suggest that the definition is theoretically valuable?
9. In terms of the DTCS's negative hypothesis:
   a. Did the definition or any of its key characteristics lead the test to faulty empirical results?
   b. Did the definition or any of its key characteristics lead the test to flawed theoretical conclusions?
   c. Does the definition obey Occam's razor; or is it a lot of work for little empirical or theoretical yield?

are made, researchers must decide which methods to use. As we have already high-lighted, choosing a methodological framework and its associates set of methods is no easy task. So, social scientists are faced with a major challenge: the DTCS requires them to test the validity of their definitions of a complex system, but such testing necessitate them to employ a new methodology, which many are not equipped to use. It is because of this challenge that, for the current project, we employed the

SACS Toolkit, which we discuss next. First, however, we need to address the final two steps of the DTCS.

Step 3: Once questions 1 through 6 have been answered, the next step is to actually conduct the test. The goal here is to evaluate the empirical validity of each of a definition's characteristics, along with the definition as a whole. In other words, along with determining the validity of each characteristic, it must be determined if the characteristics fit together. Having made that point, we recognize that not all complexity theories (particularly metaphorical ones) seek to provide comprehensive definitions; opting instead to outline the conditions and challenges, for example, that educational administrators face when coming to terms with the complexity of their organizations [51]. Nonetheless, regardless of the definition used, its criteria need to be met.

Step 4: Finally, with the analysis complete, researchers need to make their final assessment. To do so, the follow question needs to asked: In terms of the negative test found in question 9 and the null hypothesis of the DTCS, to what extent, and in what ways is (or is not) the chosen definition, along with its list of characteristics, empirically valid and theoretically valuable? With the answer to this question determined, the test is complete.

Chapter 3
Case-Based Modeling and the SACS Toolkit

Researchers in the social sciences currently employ a variety of mathematical/computational models for studying complex systems. Despite the diversity of these models, the majority can be grouped into one of four types: equation-based modeling, stochastic (statistical) modeling, computational modeling and network modeling. In the last few years, however, Byrne and colleagues have added a fifth type, called *case-based modeling* [13, 18].

Case-based modeling is an extension of the *case-based methods* tradition, which is actually an umbrella term for a variety of techniques [3, 13]. The simplest example of a case-based method is the *case study*, which medicine uses regularly. The most popular technique, however, is *case-comparative method*, which is an established approach in the social, medical and public health sciences, used for conducting in-depth, idiographic, comparative analyses of cases and their variable-based configurations [13]. Case-comparative methods vary in type and approach, from cluster analysis and discriminant function analysis to a handful of qualitative approaches, including Ragin's qualitative comparative analysis (QCA) [13].

Regardless of the approach, case-based method is defined by two key characteristics: *(1) the case is the focus of study, not the individual variables or attributes of which it is comprised; and (2) cases are treated as composites (profiles), comprised of an interdependent, interconnected set of variables, factors or attributes.*

As an example of these two points, case-based researchers would not study the isolated impact that gender or ethnicity has on the health of a neighborhood. Instead, they would study how the different profiles of neighborhoods explain their dissimilar health outcomes, with the intersecting influence of gender and ethnicity being part of the puzzle. In other words, they would view the profiles of these different neighborhoods as forming some type of 'emergent' configuration, where the whole (neighborhood) is more than the sum of its part. Each variable, therefore, would not be treated as an isolated factor impacting the neighborhoods of study; instead, each would be seen as part of a larger, context-specific set of factors, which would collectively define the case of study, and in rather complex and nonlinear ways.

This approach should make sense to public and community health scholars. Medicine, ultimately, is grounded in the case, and for good reasons: it constitutes the most effective and intuitive way to manage the complex causal structure of some

© Springer International Publishing Switzerland 2015
B. Castellani et al., *Place and Health as Complex Systems*,
SpringerBriefs in Public Health, DOI 10.1007/978-3-319-09734-3_3

set of socio-psycho-biological factors. And yet, thinking about cases is not the approach public and community health scholars employ when they go to do their research. Instead, they turn to variable-based inquiry, as defined by the majority of stochastic (statistical) methods used in public and community health research. The problem with this methodology—and this is the point the COP approach also seeks to make—is that variable-based statistics has little interest in cases or any in-depth understanding of how a set of variables collectively define or impact these cases. Instead, variable-based inquiry seeks to understand the relationship that variables have with each other, and usually in the most parsimonious, reductionist, nomothetic, linear, unidirectional manner possible. In doing so—and here is the other point of the COP approach—variable-based inquiry fails to adequately model complexity.

With this point, we move to our next section: case-based complexity science and the work of the British sociologist and complexity scientist, David Byrne.

3.1   Case-Based Complexity Science

Over the last several years, Byrne has emerged as a leading international figure in what most scholars see as two highly promising but distinct fields of study: (1) case-based method and (2) the sociological study of complex systems. An example of the former is Byrne's *Sage Handbook of Case-Based Methods*, which he co-edited with Charles Ragin, the creator of QCA. An example of the latter is his widely read *Complexity Theory and the Social Sciences*—which he just significantly updated with Callaghan in 2013 [14].

What scholars (including the current authors) are only beginning to grasp, however, is the provocative premise upon which Byrne's work in these two fields is based. His premise, while simple enough, is ground-breaking: *cases are the methodological equivalent of complex systems; or, alternatively, complex systems are cases and therefore should be studied as such.*

With this premise, which we mentioned above, Byrne introduces an entirely new approach for modeling the temporal and spatial dynamics of complex systems. Pace Byrne, *case-based complexity science* is the attempt to actively integrate case-based method with the latest developments in complexity science for the purpose of modeling complex systems as sets of cases. In turn, *case-based modeling* is the mixed-methods set of techniques scholars use to engage in case-based complexity science, the majority of which come from the computational and complexity sciences.

For Byrne (and for us) complexity scientists and case-based researchers make a similar argument: (1) variable-based inquiry is insufficient for modeling complex systems; (2) needed instead are methods that employ an idiographic approach to modeling, one grounded in the techniques of constant comparison; (3) the whole of a case or system is more than the sum of its part; (4) and yet, the study of parts and their complex interactions, from the ground-up, including the interactions these parts have with the case or system as a whole, is the basis to modeling; (5) furthermore, complex systems, as cases, introduce the notion of difference into complexity, demonstrating how profiles and their contexts lead to different traces and trajectories. We can go on with this comparison. Bottom line: cases are complex systems; complex systems are cases.

Such similarities, however, are as far as the link between case-based complexity science and the mainstream complexity sciences go. Fact is, Byrne (as well as ourselves) sees case-based complexity science as its own particular approach, distinct from other approaches currently *en vogue* in the complexity sciences. To clarify this distinction, several comments are in order.

3.2 Situating Case-Based Complexity Science

In the last thirty years, Academia has witnessed the emergence of what many scholars—including Stephen Hawking—call a 'new kind of science.' The name of this new, massively interdisciplinary science is complexity. While young, complexity science (like many new scientific innovations of late) has captured part of the academic and public imagination—in this case with discussions of six-degrees of separation, swarm behavior, computational intelligence and simulated societies. This popularity, however, has come with a price: confusion over the field's core terminology and the disciplinary divisions within it. As Mitchell explains in *Complexity: A Guided Tour*, [49] while it is popular to refer to complexity science in the singular, "neither a single science of complexity nor a single complexity theory exists yet, in spite of the many articles and books that have used these terms" (2009, p. 14).

If one follows Castellani and Hafferty, [17] however, complexity science's confusion over terminology has less to do with its age, and more to do with its interdisciplinary and therefore interstitial (between things) character. Interstitial areas of thinking, no matter how novel, replicate the dominant intellectual divisions of academia, such as science versus theory or qualitative method versus statistics. Complexity science, given that it situates itself within the full range of academic inquiry—from the humanities and the social sciences to mathematics and the natural sciences—is replete with such divisions. As such, while oriented toward the study of complex systems in general, the scholars in complexity science find themselves struggling with significant divisions regarding the complexity theories they use, the methods they employ, the epistemologies upon which they rely, and the definitions of a complex system they embrace. Given these divisions, a few clarifications are in order—all of which help us to understand better the approach of case-based complexity science.

1. The first clarification concerns the goals of science. As mentioned by Mitchell [49], complexity science is really *the complexity sciences*. To date, complexity science can be organized into several competing types, based on different combinations of the dominant distinctions in academia [14].

For Byrne (and for us), one of the most important distinctions in the complexity sciences is between what Morin [50] calls restricted versus general complexity science. *Restricted complexity science* is popular in economics and the natural sciences. It is defined as the empirical study of complex systems via the methods of rule-based, computational modeling. Its goal is quasi-reductionist, as it seeks to identify and explore the set of rules out of which complex systems emerge, so it can generate quasi-general laws about complex systems. In contrast is *general complexity science*, which is defined as the empirical study of complex systems via the methods of case-based comparative research. Its goal is more qualitative and holistic, seeking to model complex systems through a comparative analysis of cases, in order to create context-specific, grounded theoretical understandings of complex systems. Case-based complexity science situates itself in the latter approach.

As Klüver and Klüver make clear in their book *Social Understanding: On Hermeneutics, Geometrical Models and Artificial Intelligence*, [42] most sociological phenomena are simply too complex to be reduced to the emergent consequence of rule-following. A more general approach, as Byrne and Callaghan explain, [14] is one that that acknowledges this point: context and messiness and the mutual influence of macroscopic and microscopic structures and dynamics are crucial to understanding social systems.

2. The second clarification concerns computational modeling. A defining feature of the complexity sciences (restricted and general) is their reliance upon the latest developments in computational modeling. As Mitchell [49] and also Capra [16] explain, while the complexity sciences offer scholars a handful of new concepts (autopoiesis, self-organized criticality), their major advancement is method. Case in point: one can go back to the 1800s to Weber, Marx, Pareto or Spencer to find reasonably articulate theories of society as a complex system; or, one can go back to the 1950s to systems science and cybernetics (or, more recently, social network analysis in sociology) to find many of the concepts complexity scientists use today. Despite their theoretical utility—which, albeit critically received, is widespread— all the aforementioned theories ultimately stalled in terms of the study of complex systems because (amongst other reasons) they lacked a successful methodological foundation.

Computational modeling is the usage of computer-based algorithms to construct reasonably simplified models of complex systems. As discussed earlier, there are four main types of computational models used in complexity science: agent (rule-based) modeling, network (relational) modeling, stochastic (statistical) modeling, and dynamical (equation-based) modeling. Different methods yield different results. Situating itself within the latest advances in computational modeling, case-based complexity science seeks to use these tools. Byrne [15] and Uprichard, [68] for example, use cluster analysis; and our own work (as in the case of the current study) employs a long list of techniques, including agent-based modeling, cluster analysis, topographical neural nets, dynamical systems theory and complex network analysis.

But, the focus for case-based complexity is different: it is on comparing cases and searching for common case-based profiles, as concerns a particular outcome. The consequence of this focus is the causal model built—not the techniques used. Focusing on cases is a search for profiles: context dependent assemblages of factors (k dimensional vectors) that seem to explain well different types of health outcomes.

For example, one could use computational modeling to examine a set of health factors (e.g., income level, education, gender, age, and residential location) to see which case-based assemblage of these factors relate to differences in city-wide mortality rates. In short, while drawing upon the latest developments in computational and complexity science method, case-based complexity science seeks to employ a distinct mixed-methods approach grounded in the case.

3. The third clarification concerns the distinction between complexity science and complexity theory. Like complexity science, there are multiple complexity theories, which form a loosely organized set of arguments, concepts, theories and schools of thought from across the humanities and the social sciences that various scholars use in a variety of ways to address different topics [14, 17].

In terms of intellectual lineage within the social and health sciences, these theories are strongly grounded in two intersecting epistemological and theoretical traditions: the one stems from systems theory, Gestalt psychology, biological systems theory, second-order cybernetics, and ecological systems theory; while the other stems from semiotics, post-structuralism, feminism, postmodernism, constructivism, constructionism and critical realism (2).

Social complexity theories and their related epistemologies are also tied up in the substantive systems theories of sociology, anthropology, political science, economics, psychology and managerial studies. As such, complexity theories can differ dramatically from one another. For example, Niklas Luhmann uses complexity theory to articulate a new, metaphorical theory of global society (a grand theory with no agents, only a communicating society); while John Holland uses complexity theory to build a bottom-up, agent-based computational theory of complex emergent systems—see Castellani and Hafferty (2009) and Byrne and Callaghan (2013) for details.

Perhaps the sharpest distinction between complexity theory and complexity science, however, is that neither necessarily has affinity for the other. In fact, complexity theories need not be data-driven, empirically grounded, computational or scientific. They can even be anti-data, anti-empirical, anti-computation, and anti-scientific. For example, Francois Lyotard (the French theorist most famous for his usage of the term postmodernism) used early empirical research in complexity science (mainly chaos theory) to end grand narrative and place a limit on the conditions of science, which he called post-modernity. Meanwhile, most scholars in the managerial sciences use complexity theory in a prescriptive manner, with almost no empirical backing whatsoever (16). In contrast, the complexity sciences, while reliant upon key concepts from complexity theory, such as self-organization or emergence, tend to ignore theory (Mitchell 2009). For example, most restricted complexity science is theoretically vacuous.

Given the above distinctions, the generalist approach of case-based complexity science is grounded in a post-positivistic epistemology, albeit one that has learned from the errors and shortcomings of much of postmodernism and post-structuralism. This seasoned viewpoint is best described as *complex realism*, which combines Bhaskar's critical realism with Cillier's understanding that knowledge and the world are complex interdependent processes. Together, these two ideas form what Byrne calls *complex realism*. Here is an all-too-short overview of its main point. For an in-depth review, see Byrne and Callaghan (2013) and also Reed and Harvey (1996). Complex realism seeks to overcome two key problems.

The first is epistemological. Why is reality so hard to comprehend? Is it because our minds cannot know reality? No, it is not. Complex realism explains that much of the contingency in knowing (causal modeling) is not because reality cannot be apprehended. Reality escapes us because it is fundamentally complex, both in terms of the real and the actual.

Second, in relation to this complexity, we have a methodological problem as concerns existing approaches: (1) Quantitative modeling (statistics) fails us because it does not know how to model complexity and is lost in a reductionist world of variables and parsimony. (2) Qualitative modeling limits itself because it cannot deal with big data or generalization and often falls prey to problematic post-positivist ideas, such as postmodernism and radical post-structuralism. (3) Restrictive complexity limits itself because it can only deal with one type of complexity: a simplistic, ground-up emergent type, which Weaver [69] calls *disorganized complexity*. It cannot, however, deal with what Weaver [69] calls *organized complexity*, which explores the complex, qualitative interactions amongst a set of factors and their impact on some case of study. (4) And, finally, conventional case-comparative method has all the methodological tools, but it does not have yet an explicit theory of complexity and complex systems to guide its inquiry.

So, what is the solution? It is Byrne's view (and ours) that the solution is to combine complex realism, organized complexity, case-comparative method and generalist complexity theory. In turn, the goal is to use this platform to construct a mixed-methods framework for modeling complex systems, primarily by drawing upon the latest advances in computational and complexity science method. And, the link pin to this approach is the idea that cases are the methodological equivalent of complex systems. If reality and our knowledge of it is complex, then complexity is the issue to address. If complex systems are cases, then complex systems cannot be reduced to some set of rules or variables, and context has to be explicitly modeled. If cases are complex systems, then case-based researchers need a wider explicit vocabulary grounded in a wider set of methods, including computational modeling. With this basic introduction outlined, we turn now to the SACS Toolkit.

3.3 The SACS Toolkit

To address the COP's methodological challenge, we employed the SACS Toolkit: a cohesive, case-based, computationally-grounded, mixed-methods toolkit for modeling complex systems [17]. The SACS Toolkit was specifically designed to overcome the limitations that conventional methods have in modeling complex systems.

The SACS Toolkit is a variation on Byrne's [13] general premise regarding the link between cases and complex systems. For the SACS Toolkit, case-based modeling is the study of complex systems as a set of k-dimensional vectors (cases), which researchers compare and contrast, and then condense and cluster (using, primarily cutting-edge computational and complexity science methods) to create a low-dimensional model of a complex system's topography and dynamics across time/space, while preserving the complexity of the system studied.

Because the SACS Toolkit is, in part, a data-compression technique that preserves the most important aspects of a complex system's structure and dynamics over time, it works very well with databases comprised of a large number of complex, multi-dimensional, multi-level (and ultimately, longitudinal) variables. Compression, as already suggested, can be done using a variety of techniques, from qualitative to computational to equation-based.

It is important to note, however, before proceeding, that the act of data compression is different from reduction or simplification. Data compression maintains complexity, creating low-dimensional maps that can be "dimensionally inflated" as needed; reduction or simplification, in contrast, is a nomothetic technique, seeking the simplest explanation possible. This distinction is crucial. At no point during the model building process is the full complexity of a system lost. Searching for the most common case-based configurations and patterns amongst the data is a way of generating a causal model, upon which the full complexity of a topic can be arranged, managed and further data-mined.

For example, in the current study, we employed the tools of cluster analysis and topographical neural nets to cluster our 20 communities into several key groups. While these smaller groups reduce the complexity of our topic to the key health trajectories of these communities, the trajectories of each individual community are preserved—See Table 6.1 and Fig. 6.1 in the Results Section. And, as we have shown elsewhere, [55, 56] a similar preservation is possible in even big data. In other words, compression still allows us to examine every case in our database (no matter how large, complex or nonlinear); in fact, we could (and do) go on to further cluster and differentiate any one cluster into further profile gradations. It all depends upon the level of granularity sought.

In terms of the current methodological limitations of the COP approach, the other strength of the SACS Toolkit is its assembled nature [18]. While grounded in a defined mathematical framework, it is methodologically open-ended and therefore adaptable and amenable, allowing researchers to employ and bring together a wide variety of computational, mathematical, historical, qualitative and statistical methods to construct a rigorous methodological framework. Researchers can even develop and modify the SACS Toolkit for their own purposes. In short, the SACS Toolkit is an effective platform for constructing a cohesive methodological framework for modeling complexities of place and health.

The SACS Toolkit is comprised of three main components:

1. First, it is comprised of a theoretical blueprint for studying complex systems, called it social complexity theory. Social complexity theory is not a substantive theory; instead, it is a theoretical framework comprised of a series of key concepts necessary for modeling complex systems. These concepts include field of relations, network of attracting clusters, environmental forces, negotiated ordering, social practices, and so forth. Together, these concepts provide the vocabulary necessary for modeling a complex system.

2. Second, it is comprised of a set of case-based instructions for modeling complex systems from the ground up called it assemblage. Regardless of the methods or techniques used, assemblage guides researchers through a seven-step process of model building—which we review below—starting with how to frame one's topic in complex systems terms, moving on to building the initial model, then on to assembling the working model and its various maps to finally ending with the completed model.
3. Third, it is comprised of a recommend list of case-friendly modeling techniques called the *case-based toolset*. The case-based toolset capitalizes on the strengths of a wide list of techniques, using them in service of modeling complex systems as a set of cases. Our own repertoire of techniques include k-means cluster analysis, the self-organizing map neural net, Ragin's QCA, network analysis, agent-based modeling, hierarchical regression, factor analysis, grounded theory method, and historical analysis.

For detailed information on how to employ the SACS Toolkit, see Castellani and Hafferty [17] and Castellani and Rajaram [18]. Here we provide a brief overview of the modeling algorithm (Table 3.1).

Following its theoretical framework, *social complexity theory*, the database the SACS Toolkit assembles for S is comprised of two types of variables: those that make up the complex system of study—which the SACS Toolkit refers to collectively as the web of social practices W—and those regarded as environmental forces E. Together, this set of W and E form the vector configuration for each case.

It therefore follows that, because S consists of n cases $\{c_i\}_{i=1}^{n}$, and each case c_i has a vector configuration of k-dimensions, it is natural, quantitatively speaking, to represent S, at its most basic, in the form of a data matrix D as follows:

$$
D = \begin{bmatrix} c_1 \\ \vdots \\ c_n \end{bmatrix} = \begin{bmatrix} x_{11} & \cdots & x_{1k} \\ \vdots & \ddots & \vdots \\ x_{n1} & \cdots & x_{nk} \end{bmatrix}.
$$
(3.1)

In the notation above, the n rows in D represent the set of cases $\{c_i\}$ in S, and the k columns represent the measurements on some finite partition $\cup_{i=1}^{p} O_i$ of W and E—as defined in the equation below, which we have written for W:

$$(a)\ O_i \cap O_j = \emptyset\ \forall i \neq j. \tag{3.2}$$
$$(b)\ \cup_{i=1}^{p} O_i = W.$$

The same definition of the partition of W applies to E to describe a single environmental force or a collection of forces.

Based on a data mining of D—done using the techniques in the case-based toolset—the model that the SACS Toolkit creates is called the *network of attracting clusters* $(\mathcal{N})$. $\mathcal{N}$ is the actual model the SACS Toolkit creates. It is a simplification of the database studied, and is comprised of three types of maps.

Table 3.1 Variables analyzed for the 20 communities in the summit county database

Compositional factors	Population 65 years of age of older[a]
	% White Population[a] (Defined as number of persons identifying themselves as "White" in response to the 1990 US Census or "White Alone" in response to the 2000 US Census)
	% African-American Population[a] (Defined as the number of persons identifying themselves as "Black or African-American" in response to the 1990 US Census or "Black or African-American Alone" in response to the 2000 US Census)
	Median Household Income[a]
Contextual factors	Overall Poverty[a] (Defined as the number of persons living "below the poverty level" as defined by the U.S. Census)
	Public Assistance[a] (Defined as the number of households receive public assistance as defined by the U.S. Census)
	Persons 25+ Years with High School Diploma[b]
	Net Job Growth[c] (Defined as the number of jobs in 2000 minus the number of jobs in 1990
	Unemployment Rate[a] (Defined as unemployed civilian labor force)
	Housing affordability[a] (Defined as the percentage of households where mortgage/rent is greater than 30 % of the household income)
	No Health Care Coverage[d] (An estimate of the number of individuals with no health care coverage based upon a statewide survey (Behavior Risk Factor Surveillance Survey–Centers for Disease Control and Prevention)
Health outcomes	No First Trimester Prenatal Care[d] (Defined as the number of births occurring to mothers from 1995 to and including 1998 for which no prenatal care was received during the first three months of the pregnancy)
	Teen Birth Rate[d] Defined as the number of births occurring between 1995 and 1998 to mothers 15 to and including 17 years of age)
	Childhood Immunization Rate[e] (Defined as the percentage of children with a complete immunization series 4:3:1 by their second birthday based on the kindergarten retrospective study)
	Child Abuse/Neglect[f] (Defined as the number of referrals resulting in assessment per 1,000 childre under 18 years of age)
	Elder Abuse/Neglect[g] (Defined as the number of referrals received by the Department of Jobs and Family Services for abuse, exploitation, or neglect)
	Years of Potential Life Lost per Death[e] (Defined as the sum of the differences between the age at death and the life expectancy at age of death for each death occurring between 1990 and 1998 due to all causes divided by the number of deaths due to all causes within the census tract cluster borders where those borders are defined by United States Census Bureau census tracts)

[a] United States Census Bureau 1990 and 2000 Decennial Censuses
[b] Ohio Department of Education
[c] NODIS
[d] Akron City Health Department, Office of Epidemiology
[f] Ohio Department of Health
[g] Children's Services Board
[h] Summit County Department of Jobs and Family Service

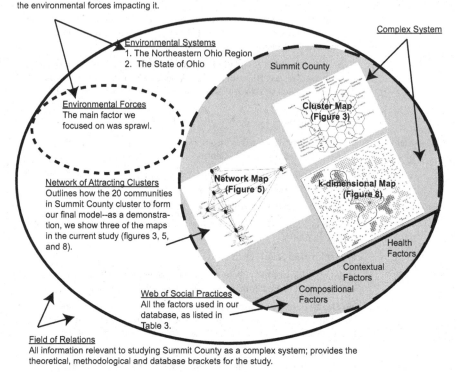

Envrionmental Systems and Forces
This outer circle contains all information relevant to the
environmental systems in which Summit County is situated; and
the environmental forces impacting it.

Fig. 3.1 Example of the final map created by the SACS toolkit for current case study

1. *Cluster Maps*: Given that the SACS Toolkit studies S as a set of cases $\{c_i\}$, it is necessary, at some point, particularly in the case of large databases, to identify and map the most common vector configurations or trajectories in S. To do so, we employ a variety of methods, including cluster analysis, topographical neural nets, genetic algorithms, ordinary differential equations and partial differential equations, specifically the advection equation. [55, 56] The results of such analyses are the cluster maps for $\mathcal{N}$, which can be static or longitudinal, and discrete or continuous. In the current study, Table 6.1 and Fig. 6.1 are static/discrete examples of such maps.

2. *Network Maps*: In turn, network maps compress S to model the most important relationships (ties, links, etc) and interactions that exist amongst its cases $\{c_i\}$, particularly as they relate to the most common vector configurations in S. To do so, we employ a variety of techniques from the field of social and complex network analysis. In the current study, Fig. 7.1 is an example of such a map.

3. *K-dimensional Maps*: The purpose of the k-dimensional maps is to understand how the variables (W, E) comprising the vector configurations for $\{c_i\}$ influence

the structure and dynamics of S. To do so, a wide list of techniques can be used, from geospatial modeling and factor analysis to logit-probit models and survival curves. In the current study, examples of such maps include Table 5.1 and Fig. 6.2.

With its set of maps constructed, the next step in the assemblage algorithm is to explore how the maps for $\mathcal{N}$ inform one another, with the goal of arriving at a well-developed, albeit simplified, low-dimensional model of S for one discrete moment in time/space. Figure 3.1 is one example of what such a model looks like (for one discrete moment across time/space) for our study of Summit County. During the process of model building and running our litmus test on Summit County, we would use this map as a reference point and as a basis for organizing our results, discussions and conclusions.

3.4 Modeling the Temporal/Spatial Dynamics of Complex Systems: A Case-Based Density Approach

While important, constructing a static model of $\mathcal{N}$ is not where the SACS Toolkit necessarily ends the assemblage process. For the SACS Toolkit, cases $\{c_i\}$ are ultimately dynamic and evolving, both temporally and spatially.

We have, therefore, been developing a novel approach to modeling aggregate sets of cases across both discrete and continuous time/space. We refer to this technique as a *case-based density approach*. The mathematical details of this approach (both in discrete and continuous time/space) are far too complicated to address in the current paper, and have been dealt with extensively in Castellani and Rajaram (2012) and Rajaram and Castellani (2013, 2014). But, we can say a bit here.

In terms of the current study, we employed a discrete case-base density approach, examining change between two time-points: 1990 and 2000. However, in Rajaram and Castellani (2013) we devoted an entire study to a continuous case-based density approach for the same case study—modeling our 20 communities from Summit County continuously and longitudinally.

To conduct our current discrete analysis, we treated the communities (cases) $\{c_i\}$ in Summit County as discrete dynamical systems $c_i(j)$, where j denotes the time instant t_j. In so doing, our goal was to model the case-based structure and dynamics of S (in this case, Summit County) as it evolves across time/space. To do so, we used the SACS Toolkit's *case-based tool set* to generate an $\mathcal{N}$ for each pf the two moments across time/space studied. The result of these repeated assemblages of $\mathcal{N}$ was a two time-stamp series of discrete, simplified, low-dimensional models of S that have both dynamical and topographical features. As a final step in the assemblage algorithm, we used the SACS Toolkit to assemble this series of models to form its final case-based model of S. See tests 2, 4, 6, 7 and 9 (all chapters in the Results Section of this study) as they deal (in varying degrees) with the issue of dynamics across time/space.

In Rajaram and Castellani (2013, 2014), we employed a different set of tools to model the continuous, longitudinal dynamics of Summit County—based on the work of Rajaram (the second author of the current study) and his colleagues, who have developed a new stability theory of ordinary differential equations called *almost everywhere uniform stability*, [54, 57, 58]. This approach—which makes use of cluster analysis, topographical neural nets, genetic algorithms, ordinary differential equations and the advection equation—is useful for modeling the temporal/spatial dynamics of social complexity in big-data in several important ways. It allows researchers to: (1) employ multiple methods; (2) map the complex, nonlinear evolution of ensembles (or densities) of cases; (3) classify major and minor clusters and time-trends; (4) identify dynamical states, such as attractor points; (5) plot the speed of cases along different states; (6) detect the non-equilibrium clustering of case trajectories during key transient times; (7) construct multiple models to fit novel data; and (8) predict future time-trends and dynamical states. This, then, is a summary of our method.

Chapter 4
Methods

Now that we have a basic sense of the Definitional Test of Complex Systems (DTCS) and the SACS Toolkit, we turn to a brief overview of: (1) our case study, Summit County and its 20 communities, (2) the specific measures we will use to construct the profiles (k-dimensional vectors) for our 20 cases, and (3) the case-based tools we will employ to conduct our litmus test.

4.1 Summit County and its 20 Communities

The case study for this paper is Summit County, Ohio, USA and its 20 communities. Summit County is typical of many Midwestern communities in the United States: it contains a major city (Akron) struggling to survive globalization and the shift from industrialism to post-industrialism, along with first-tier and second-tier suburbs, as well as a few semi-rural communities. In terms of wealth, there are very poor, urban communities with poverty traps (e.g., Southwest Akron); first-tier suburban, middle-class communities (e.g., Cuyahoga Falls); and very affluent suburban communities, such as Hudson—see Fig. 4.1. One finds the typical American health disparities in this county as well. For example, while the average "years of life lost per death" for affluent Hudson is 10.5; it is 17.4 for Southwest Akron. In short, there is nothing much anomalous in this county in comparison to most of the American, Midwestern places studied in the community health science literature.

Our identification of the 20 communities in Summit County was based on census tracts data. Summit County is comprised of 121 census tracts. Public health researchers in Summit County created the analytical boundaries of these 20 communities by clustering census tract data according to identifiable communities, cities, towns, neighborhoods and ethnic groupings. Tracts were also clustered to maintain demographic homogeneity within the 20 communities. For more information, see the Healthy Summit 2000 Health Indicators Summary Report [67].

© Springer International Publishing Switzerland 2015
B. Castellani et al., *Place and Health as Complex Systems,*
SpringerBriefs in Public Health, DOI 10.1007/978-3-319-09734-3_4

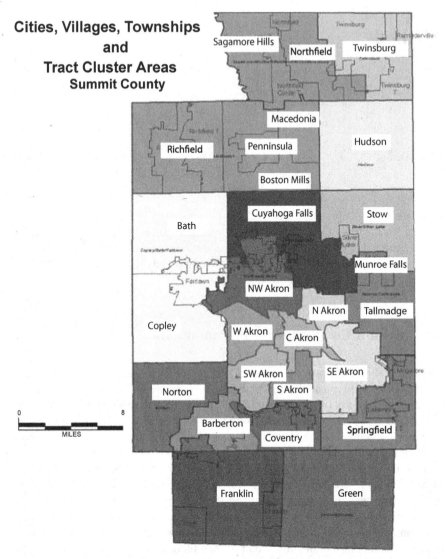

Cities, Villages, Townships and Tract Cluster Areas
Summit County

Sagamore Hills

Northfield

Twinsburg

Macedonia

Hudson

Richfield

Penninsula

Boston Mills

Cuyahoga Falls

Stow

Bath

Munroe Falls

NW Akron

N Akron

Tallmadge

Copley

W Akron

C Akron

SW Akron

SE Akron

Norton

S Akron

Barberton

Coventry

Springfield

Franklin

Green

0   8

MILES

Source: This map was retrieved from http://www.healthysummit.org/Summit2020QoL
.html on the 29th of November, 2011. It is a public document provided by the Health
Summit 2010 website. SOURCE: Federation for Community Planningvand
Northern Ohio Delta & Information Services (NODIS) Maxine Goodman Levin
College of Urban Affairs, Cleveland State University, USA, April 2003.

Fig. 4.1 Summit county and its 20 communities

4.2 Database

To conduct our tests we partitioned a database from the *Summit 2010: Quality of Life Project* (Summit-QLP). The Summit-QLP is a website that houses twenty years' worth of information on the health and wellbeing of Summit County. All of the data are in the form of PDF reports, providing detailed statistical and qualitative information on Summit County and its 20 communities. To obtain our measures, we combed through the statistical reports, entering data for the 20 communities into a database for 1990 and 2000—the two points in time during which Summit-QLP collected data. We used both time points in our case study because, as we identified above, sprawl is about the evolution of places, and so longitudinal data is important. Similarly, the ultimate goal of complexity science is to study how complex systems change over time. Given the constraints of data collection, however, some of the health outcomes data only represent one point in time—see Table 3.1 for explanation.

4.3 Measures

As shown in Table 3.1, the measures provided in the *Summit 2010: Quality of Life Project* (Summit-QLP) constitute a rather conventional set of measures, which can be grouped into one of three types: compositional factors, contextual factors and health outcomes.

Compositional Factors Two of the most commonly used measures of composition are median household income and ethnicity [47]. When combined and examined across time, they can also be used as indirect measures of sprawl: in other words, across time, household income and ethnicity allow us to track how the overall compositions of communities changed, particularly "white affluent flight" into the suburbs—See Table 3.1.

Contextual Factors As shown in Table 3.1, the contextual measures from Summit-QLP, expressed as rates, address a variety of key sociological and economic factors, including economy (job growth, civilian labor force, poverty and unemployment), housing (mortgage/rent to income ratio), education (high school completion) and health care (health insurance and public assistance).

Health Outcomes Sprawl and the community-level segregation of wellbeing are linked to a variety of health outcomes in the literature, in particular early warning measures (e.g., birth weight), adult health measures (e.g., hypertension), mental health (e.g., stress and wellbeing) and mortality (e.g., [9, 28]). For our study, to track the impact of sprawl on community-level health, we employed the following measures from the Summit-QLP: 1st trimester care, childhood immunizations (early life); teen pregnancies (adolescent health); child and elder abuse (mental health); and years of life lost per death (adult health and mortality)—See Table 3.1.

4.3.1 Case-Based Toolset

While the SACS Toolkit provides a cohesive, mixed-methods platform for modeling the temporal and spatial dynamics of complex systems, not all of its case-based tools need be used for a given study. Our goal here is to review the tools we used to conduct the litmus test for the current study, focusing on those most likely to be new or novel to most readers. First, to generate our cluster maps (e.g., Table 6.1 and Fig. 6.1) we used k-means cluster analysis (abbreviated k-means) and the Self-Organizing Map (SOM) [43]. For k-means, we used the statistical software package SPSS. For the SOM we used the SOM Toolbox (www.cis.hut.fi/projects/somtoolbox/download/). The SOM Toolbox runs as a function package in the MATLAB computing environment. Second, to generate our network map we used the freeware Pajek (http://vlado.fmf.uni-lj.si/pub/networks/pajek/). Our k dimensional maps (e.g., Figs. 6.2, 7.1, 4.2, 13.2) were generated using SPSS, the SOM, and agent-based modeling. Finally, to generate our agent-based model for our k-dimensional map, we built and ran our agent-based model (*called Summit-Sim*). We built Summit-Sit in NetLogo, a freeware program (ccl.northwestern.edu/netlogo/).

SOM and K-Means: We used k-means and the SOM to do three things: construct a causal model of the profiles for each of our 20 cases (communities), cluster these cases into the most salient profiles, and then model the change and evolution of these profiles across time/space. K-means is a partitional (as opposed to hierarchical), iterative clustering technique that seeks a single, simultaneous clustering solution for some proximity matrix [37]. K-means is also a form of unsupervised learning: unlike classification techniques, the cluster member of a case is not known prior to analysis. The SOM is part of the distributed artificial neural network literature [43]. In this literature, the SOM serves a specific function: mapping high-dimensional data onto a smaller, two-dimensional space, while preserving, as much as possible, the complex, non-obvious patterns of relationships amongst this data [43]. The SOM's strength is its capacity to generate rich, visually intuitive clusters.

The ultimate strength of these two techniques is that they work well together [44]. For example, k-means and the SOM have a similar approach to data compression and clustering. Both can be viewed as vector quantization techniques, insomuch as they cluster cases by searching for a simpler set of reference vectors, with each case in $\{c_i\}$ being positioned near its most similar reference vector. For k-means, the reference vector is a centroid, which represents the average for all the cases in a cluster. For the SOM, the reference vector is an actual point, a neuron, which represents the weighted average of the vector configurations clustering around it. From here, however, they differ. But, it is their differences that make them work so well together.

K-means is useful because it requires the number of centroids to be identified ahead of time, based largely on theoretical rationales. Such an approach is important for case-based modeling because it requires that the selection of clusters be somehow theoretically driven—researchers should have some sense of what they are looking for, based on a preliminary study and comparison of the cases $\{c_i\}$ in S.

In turn, the SOM functions as an effective method of validating k-means because the set of reference vectors (neurons) it settles upon is not predetermined. If, therefore, the SOM arrives at a solution similar to the k-means, it provides an effective method of corroboration. The SOM is also useful because it graphs its reference vectors and the cases $\{c_i\}$ surrounding them as neurons on a two-dimensional, topographical map, called the U-matrix. On the U-matrix, the reference vectors most like one another are graphically positioned as nearby neighbors, with the most unlike reference vectors (neurons) being placed the furthest apart. The U-matrix therefore provides a visually intuitive, low-dimensional map of the original high-dimensional database being studied—See Fig. 6.1.

The rationale for using these two techniques to model the discrete (two time stamps) evolution of a network of communities across time/space is as follows: If, following case-based modeling, complex systems typically contain a large number of cases, and if the vector configurations for each case $\{c_i\}$ in a complex system S generally share common profiles (both in terms of proximity and adjacency), then a useful method of longitudinally modeling S, according to the SACS Toolkit, is to cluster it. Clustering is effective because it allows for the identification, mapping and analysis of the most common vector configurations in S for each discrete moment in time/space. This discrete network of attracting clusters can then be treated and mapped as the trajectories (attractor points) for a network of communities across time/space.

Agent-based Modeling: As we discussed in the introduction, the complex systems definition used by the COP approach contains nine key characteristics (Table 1.1). To test Characteristic 9 (C9), we employed the tools of agent-based modeling.

Agent-based modeling is a bottom-up approach to simulating complex systems. It is based on the viewpoint that many social outcomes emerge from the micro-level interactions amongst a heterogeneous set of rule-following agents, as they take place across time [31]. The theoretical rationale for using agent-based modeling in the current study is as follows: if sprawl is a complex systems phenomenon that emerges out of the interplay between compositional and contextual factors, then agent-based modeling is necessary, as conventional methods cannot model such an interaction across time.

Summit-Sim: The model built for the current study is called Summit-Sim—See Fig. 4.2. Summit-Sim is a rule-based, multi-agent, discrete-event model with a 51X51 lattice structure, upon which a randomly distributed set of upwardly-mobile agents migrate to find their ideal residential location. Summit-Sim was designed to explore the link between residential migration patterns and health outcomes, based on our empirical study of Summit County and its 20 communities.

For the current study, a simplified version of Summit-Sim was used. Readers can run or downloaded (including code) at cch.ashtabula.kent.edu/summitsim.html. This simple version looks at only one aspect of sprawl: how the residential migration patterns of upwardly mobile agents influence community-level health outcomes. Later developments of this model will incorporate the interplay between residential migration and macroscopic factors such as schooling and job growth. See results section for more details on our theoretical focus.

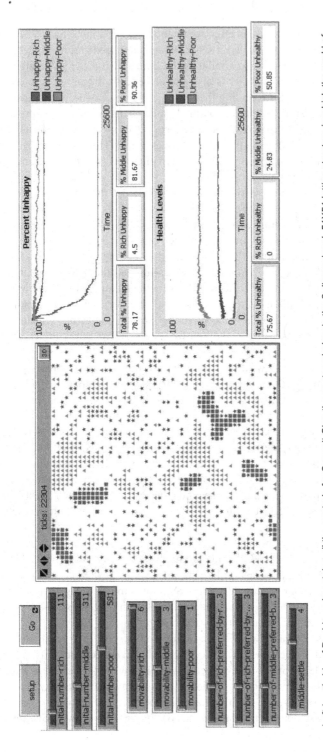

Left-hand side of Dashboard contains all the controls for Summit-Sim; the center window is the 2-dimensional, 51X51 lattice structure in which the world of Summit-Sim takes place; while the right-hand side contains the two output graphs for happiness ratings and healthiness ratings.

Fig. 4.2 Snapshot of the Netlogo Dashboard for Summit-Sim

Figure 4.2 provides a snapshot of the dashboard for Summit-Sim. On the dashboard are three types of information: the world of Summit-Sim (in the middle); Controls (on the left); and Output Charts (on the right).

Three types of agents inhabit the world of Summit-Sim: rich agents (squares), middle-class agents (triangles) and poor agents (circles)—the number of which is determined by the three population sliders on the left-side of the Summit-Sim dashboard. To test C9 (characteristic 9) we used several factors from our study of Summit County to determine the population of agents for our simulation: differences in household income (circa 1999/2000); education; the capacity to change residential location; and work type. Poor Agents represent those agents in Summit County who are financially struggling. In 2000, the median household income in Summit County was 42,000. Using Census data, poor agents represent roughly 58 % of Summit County (with a household income range of 0 to 49,000). Poor agents are associate-level educated or less, with work ranging from low-level white collar work to unemployment. Despite differences, all poor agents have difficulty changing residential location given their financial situation. Middle class agents (31 % of Summit County) make between 41,000 and 99,000, have some college education or higher, work skilled-blue collar, white-collar or professional jobs, and are moderately able to change residential location. Affluent agents (11 % of Summit population) represent those households in Summit County making 100,000 or more, who are generally college educated, have professional-class jobs or lucrative blue-collar jobs, and are able to change easily residential location.

In terms of controls, three rules govern the discrete migration behavior of Summit-Sim agents. These rules are Preference, Preference-Degree, and Mobility.

Preference is a modification of Schelling's well-known segregation rule. [10] Unlike the original Schelling model, however, wherein agents seek to live near their own kind, Preference concerns sprawling, upwardly mobile agents migrating to live near agents of a similar or higher status. While sprawl produces segregated neighborhood, it is not necessarily about agents migrating to live near similar agents. Sprawl is about agents migrating to live in better neighborhoods. For rich agents, 'better' means neighborhoods with more rich agents. For middle agents, 'better' means living near more rich agents, or at least lots of middle-class agents. For poor agents, 'better' means living near middle-class agents, if they can. Following this logic, in Summit-Sim, at each discrete point in time, (a) rich agents seek to live near rich agents; (b) middle-class agents seek to live near rich agents; if they cannot, they seek to live near other middle-class agents; if they find themselves in a neighborhood with 4 or more middle-agents, they stay; and (c) poor agents seek to live near middle-class agents; if they cannot, they stay where they are.

Preference-Degree determines the number of higher status agents around which agents prefer to live. In a 2-D lattice structure, 'neighbors' is defined as the total number of spaces (squares) available around an individual agent, which range from 0 to 8. In Summit-Sim, preference ranges from 1 to 3. Our more exhaustive tests find that, if preference is set beyond 3, the model is unable to settle.

Movability is the capacity for an agent to migrate to the neighborhood in which they ultimately desire to live. Ranging from 1 to 6, mobility is defined as the number of spaces an agent can move per iteration. Following our empirical analysis of Summit County, we set the movability of poor agents at 1, primarily because it is very difficult for these agents to buy homes, sell homes or rent a more expensive apartment in order to move. We set the movability for middle-class agents at 3, because they are moderately able to sell their homes or buy a new home or rent a more expensive apartment in order to change location. And, we set the movability at 6 for rich agents because they can move with little effort.

Summit-Sim is comprised of two major charts. The first chart is an unhappiness rating. At each iteration agents are asked if they are happy. Happiness is defined as living in the type of neighborhood they seek. The Unhappiness Chart maps the percentage of affluent, middle-class and struggling agents unhappy after each iteration.

The second chart is a healthiness rating. The COP literature generally has shown that residential segregation tends to produce health inequalities insomuch as the less affluent individuals there are in a community, the worse its health outcomes, due in large measure to the complex interplay between compositional factors, such as household income and the health of local institutions, such as schools [8, 22, 60]. Following this argument, we used a rough context-based indicator of community-health. First, we began each simulation of our model with the health of all agents (poor, middle class and rich) being equal. If, however, once the model was started, the 9×9 region in which an agent was living had three or more rich agents, they were considered healthy. At the aggregate level, we were then able to express the healthiness of our three groups as a percentage: overall health, followed by percentages for rich, middle and poor agents.

Results

In this section, we will proceed to test each of the COP's nine characteristics, one at a time, in litmus test fashion, to examine the theoretical and methodological utility of the COP approach, as applied to the topic of sprawl. For each characteristic, we summarize what the COP literature has to say about it, followed by a discussion of the methods we used to test the characteristics and then our results. As a final note, from here on out, as we stated in the Introduction, we will refer to these characteristics, in their entirety, as the COP's 9-characteristic definition; also, we will refer to each characteristic in abbreviated form—for example, Characteristic 9 (the idea that places are agent-based) will be labeled C9.

Chapter 5
Places Are Complex

In terms of studying sprawl and health, a major issue is modeling the complex causal relationship between compositional and contextual factors. We therefore begin with Characteristic 1. The theme of $C1$ takes two forms. In its positive form, the main point of $C1$, as Cummins et al. [21] state, is that "research in place and health should avoid the false dualism of context and composition by recognizing that there is a mutual reinforcing and reciprocal relationship between people and place" (p. 1825). In its critical form, the main point of $C1$ is that the "tight interrelationships between individual [composition] and context are not easy to capture in quantitative studies" (Cummins et al. p. 1829). "This is partly why," explain Cummins et al. (2007, p. 1829), "some researchers have adopted important alternative methodological strategies such as qualitative techniques."

To test $C1$, we used linear modeling to see if we could (according to convention) parse the independent contribution that the compositional and contextual factors listed in Table 3.1 have with two of our study's health outcomes: Years of Potential Life Lost per Death (YLL) and Teen Birth Rate (TBR). Two notes: (1) the first outcome was chosen because it is a widely used global measure of community-level health; the second was chosen randomly, so as to report on more than one outcome; however, results similar to those reported below were found when we examined the other outcomes; (2) we analyzed a single time frame for our health outcomes because they are single measures for multiple years, YLL (1990–1998) and TBR (1995–1998)—See Table 3.1 for more information.

For our linear modeling we used zero-order correlation first, and then hierarchical regression. We ran zero-order first because it shows the bivariate (pairwise) correlation between each factor and our two outcomes, ignoring the statistical influence that other factors have on the relationship. We ran hierarchical regression second because such analyses demonstrate the "independent" relationship that compositional and contextual factors have with our health outcomes, after the effects of the other factors have been statistically removed from the equation. If, after running hierarchical regression, either set of factors (compositional or contextual) have roughly the same relationship with our two health outcomes (after controlling for the other set of factors) it would support the hypothesis that compositional and contextual factors

© Springer International Publishing Switzerland 2015
B. Castellani et al., *Place and Health as Complex Systems*,
SpringerBriefs in Public Health, DOI 10.1007/978-3-319-09734-3_5

Table 5.1 Conventional analysis of compositional factors, contextual factors and health outcomes

	Variable	Column 1 Zero-order correlations		Column 2 Hierarchical regression	
		Years of life lost per death	Teen birth rate	Years of life lost per death	Teen birth rate
Composit-ional factors	1990 % pop (65year&older)	0.048[a]	0.435	ns	ns
		(0.840)[b]	(0.056)		
	1990 % non-hispanic caucasian in 1990	−0.541	−0.781	−0.672[c]	ns
		(0.014)	(0.000)	(0.008)	
	% African-American in 1990	0.551	0.768	ns	ns
		(0.012)	(0.000)		
	1990 household income	−0.794	−0.814	ns	ns
		(.000)	(.000)		
Contextual factors	% overall poverty 1990	0.636	0.926	ns	ns
		(0.003)	(0.000)		
	% public help 1990	0.682	0.944	ns	ns
		(0.001)	(0.000)		
	% no high school 25years+ in 1990	0.752	0.809	ns	ns
		(0.000)	(0.000)		
	Job growth (1993 to 2000)	−0.204	−0.493	ns	ns
		(0.387)	(0.027)		
	% unem-ployed 1990 .671	(0.001)	(0.000)	ns	0.702
		(0.001)	(0.000)		(0.001)
	% of households mortgage/rent is <30 % of income	0.532	0.912	ns	ns
		(0.016)	(0.000)		
	% no health care coverage	0.684	0.939	ns	ns
		(0.001)	(0.000)		

Column 1 provides zero-order, pairwise correlations for all compositional and contextual factors listed in Table 3.1 with two health outcomes: years of life lost per death and Teen Birth Rate. In this column, [a] is the correlation coefficient; and, [b] is its two-tailed, significance level.

Column 2 provides the results of our hierarchical analysis of the "independent" relationships all compositional and contextual factors listed in Table 3.1 with two health outcomes: years of life lost per death and Teen Birth Rate. In this column (ns) is a non significant partial correlation coefficient; [c] is a significant partial correlation coefficient for a two-tailed significance level

are independent. Table 5.1 shows our results: no such "independent" relationship was found. Let us explain.

Column 1 in Table 5.1 lists the zero-order correlations between our compositional and contextual factors and our two health outcomes. Three things stand out in Column 1. First, YLL and TBR correlate significantly with almost every factor. Second, whatever 'direction of relationship' a compositional or contextual factor had with one health outcome, it had with the other. For example, unemployment in 1990 correlates positively and significantly with both TBR and YLL. Third, several compositional and contextual factors were highly correlated with our two health outcomes. For example, 11 of the 22 correlations were at 0.75 or better.

Our hierarchical analysis, shown in Column 2, however, reveals a different picture. The only factor that had any remaining impact on YLL was the percentage of the population that was non-Hispanic Caucasian (partial correlation coefficient $= -0.672; p = 0.008$). No contextual variables (shown in Column 2 as ns, not significant) retained significance when predicting YLL. Conversely, the only factor that had any remaining impact on TBR was the percentage of unemployed (partial correlation coefficient $= 0.702; p = 0.001$); and no compositional variables remained significant when controlling for contextual variables.

Complex linear models rely on the ability to observe predicted patterns. In the current example we see strong zero-order associations, but the associations do not maintain predicted patterns when statistical control is exerted on theoretically relevant variables. This inconsistency may, as the positive form of $C1$ suggests, reflect an inaccurate pattern of predicted associations; that is, compositional and contextual variables are not independent of each other. More likely, given the high level of multicollinearity between our factors, it is likely that our results reflect statistical anomalies. Either way, linear modeling does not provide much information.

Chapter 6
Places Are Emergent and Self-Organizing

In terms of its theme, $C2$ begins where $C1$ ends. If linear statistics cannot adequately model the relationship between health outcomes and compositional and contextual factors, how should researchers model communities? According to $C2$, communities should be modeled as case-based, complex configurations that emerge out of the self-organizing interactions amongst a set of compositional and contextual factors and their related health outcomes.

By emergence, these scholars mean that a community's resulting configuration is such that (a) the whole is more than the sum of its compositional, contextual and health outcomes parts, and (b) one cannot understand this whole through reductionism; the community must be understood as a system [36]. For example, as Curtis and Riva state: "Complexity theory also anticipates that health systems are dynamic and have an inbuilt capacity to organize and reorganize themselves constantly (emergence and re-emergence of human diseases being an illustration)" ([24], p. 2)

As Gatrell [30] explains, by self-organization they mean that the configuration (order) that emerges out of the intersection of a set of compositional and contextual variables and their related health outcomes is: (a) self-sustaining in the face of environmental pressures; (b) not the direct result of any *a priori* design on the part of community, etc; (c) not strictly determined or controlled by any one internal or external supervisor or force; and (d) is more than just the processes of internal feedback loops that can be explained via a linear model ([19], pp. 90–93).

As Blackman [7] explains, by case-based (as discussed in our methods section), they mean two things. First, they mean that places are cases. For these scholars, places are cases in the idiographic sense that each place constitutes a unique configuration of compositional and contextual factors. As Rihoux and Ragin explain [59], a place is a "complex combination of properties, a specific 'whole' that should not be lost or obscured in the course of the analysis—this is a holistic perspective" (p. 6). Second, they mean that, given this holistic view, the study of place cannot be variable-based. Instead, it needs to be case-based, where places are treated as holistic composites of a set of interacting variables and their changes over time. As Blackman [7] states, "complexity theory focuses on cases as empirical and actual domains. The interest is in the 'states' of these cases; not so much how and why variables change but how and why cases change" (p. 31).

© Springer International Publishing Switzerland 2015
B. Castellani et al., *Place and Health as Complex Systems*,
SpringerBriefs in Public Health, DOI 10.1007/978-3-319-09734-3_6

Our analysis of Summit County using the SACS Toolkit suggests that $C2$ is an empirically valid and theoretically valuable way to (a) understand the complex causality of places and (b) address the limitations of using conventional method to determine such a relationship.

To test $C2$, we employed two case-based comparative techniques: the self-organizing map algorithm and k-means cluster analysis [17]. As explained in our methods section, we employed these techniques for two reasons. First, they are empirically-driven, iterative techniques designed to explore the configurations that emerge out of the clustered self-organization of a given set of cases and the variables upon which these cases are based—in fact the SOM is literally named the self-organizing map. Second, given the instability and inconsistency often associated with cluster analysis, these techniques corroborate one another [4, 44, 26].

We had two goals for our test of $C2$. Using the SACS Toolkit, our first was to determine if there was any order in the idea that communities are complex configurations that emerge out of the self-organizing interactions amongst a set of compositional and contextual factors and their related health outcomes. To do so, we treated the communities in Summit County as 20 separate cases, each representing a different configuration of the compositional, contextual and health outcome variables in our study—see Table 3.1. We began with k-means cluster analysis. We started with k-means because it allowed us to assign our 20 cases to a fixed number of clusters so we could explore different cluster solutions. We also used k-means because it creates single-rank clusters.

For our analyses we used all 17 variables listed in Table 3.1. We explored normalizing household income because its range was greater than the other variables, but it did not improve the results. We also ran our equation altering variable entry to control for any ordering effect. Finally, we ran the k-means with different fixed cluster solutions. The final solution, shown in Table 6.1 was a seven cluster solution. We settled on this solution because it separated the cases well without lumping them into any one cluster. Also, it fit with our expert knowledge of Summit County: as suggested by the literature, one effective way to manage the instability of cluster analysis is to have (if possible) prior knowledge about how one's cases should cluster [1].

Here is our breakdown of the clusters. First, there are the affluent suburban communities, which include Hudson (Cluster 4) and Copley Bath and Fairlawn (Cluster 5). Of the two clusters, Hudson is the richest and significantly differs from all other 19 communities, particularly in terms of health outcomes. Next, there are the middle class suburban communities, which include Stow/ Silverlake, Northfield/Macedonia/Sagamore, and Richfield/Peninsula (all in cluster 1). These communities are followed by the middle to working class suburbs and semi-rural areas, which include Springfield, Coventry/Green and Cuyahoga Falls (Cluster 6) and Twinsburg, Northwest Akron, Munroe Falls/Tallmadge, Norton and Franklin (Cluster 3). Of the two, Cluster 3 has a slightly higher average household income and a larger African-American community. Finally, there are the poor inner-city communities, which include all of the communities in the city of Akron (except northwest Akron) as well as the city of Barberton—all of these communities are in Clusters 2 and 7. Cluster 2 has one community, the poorest in Summit County, Central Akron.

Table 6.1 Final K-means Cluster solution for 20 communities in summit county

Variables (Unless otherwise noted, all data is from 1990—See Table 6.2)	Cluster						
	1	2	3	4	5	6	7
% Non-hispanic caucasian	97.3[a]	68.6	93.5	97.6	93.8	98.4	77.5
% African-American	1.7	28.0	5.6	1.0	4.7	1.0	21.2
% Overall poverty	3.60	44.30	6.04	1.00	2.60	6.77	19.30
1990 household income	41,464	11,404	36,021	68,083	49,144	30,002	21,688
Job growth (1993–2000)	31.87	20.80	17.36	27.70	43.10	15.83	0.33
% Civilian Labor Force (16+ old)	96.17	85.90	95.22	96.60	95.70	94.73	90.82
% Receiving public assistance	2.8	25.8	4.3	1.4	2.6	5.6	13.8
% No high school degree (25 year+)	15.3	41.5	16.8	2.7	11.1	22.1	29.4
% of households mortgage/rent is <30 % of income	16.0	43.4	17.6	15.8	19.0	18.1	27.4
% Unemployed	3.8	14.1	4.8	3.4	4.3	5.3	9.2
% No 1st trimester care 1995–1998	5.63	24.60	7.54	1.20	4.80	8.90	14.78
Teen pregnancies per 1000 births (1995–1998)	5.80	66.00	12.54	1.30	3.50	12.33	47.72
% Children immunized by 2 year of age	74.1	40.0	76.5	86.1	72.9	78.1	60.7
% No Health Care Coverage	4.20	25.30	6.34	1.20	3.70	8.40	14.52
Child abuse/neglect rate per 1000	10.8	98.3	19.3	4.0	6.8	16.2	60.5
Elder abuse/neglect rate per 1000	4.1	53.8	4.9	2.1	4.8	9.1	9.3
Years lost per death 1998	13.83	16.40	13.96	10.50	10.60	14.40	15.18

[a]The values listed in the columns for all 7 clusters represent the average value/measurement that the communities in that cluster scored for each variable listed in Column 1. In cluster analysis, these averages are called the cluster's centroids. **2. Community Membership for each of the 7 Clusters is as follows: Cluster 1**: Stow/ Silverlake, Northfield/Macedonia/Sagamore, and Richfield/Peninsula; **Cluster 2**: Central Akron; **Cluster 3**: Twinsburg, Northwest Akron, Munroe Falls/Tallmadge, Norton and Franklin; **Cluster 4**: Hudson; **Cluster 5**: Copley/Bath/Fairlawn; **Cluster 6**: Springfield, Coventry/Green and Cuyahoga Falls; **Cluster 7**: North, West, Southwest, South and Southeast Akron and Barberton City

With our initial cluster solutions determined, we proceeded to corroborate our k-means with the SOM. The SOM functions as an effective method for validating k-means because the set of reference vectors (neurons) it settles upon is unsupervised. If, therefore, the SOM arrives at a solution similar to the k-means, it provides an effective method of corroboration. The SOM is also useful because, as Fig. 6.1 shows, it spatially graphs its reference vectors (similar to k-means centroids) and the cases (c_i) surrounding them onto a variety of n-dimensional surfaces. The 2-dimensional

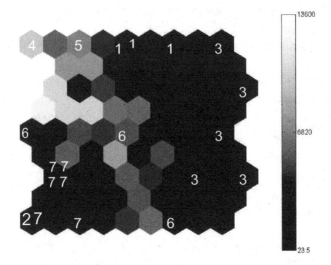

(Topographical Solution for Case Study. Index for 2a on right is unstandardized
distance, moving from low values (dark) to high values (light) The lighter the polygons
the greater their conceptual distance is from one another.)

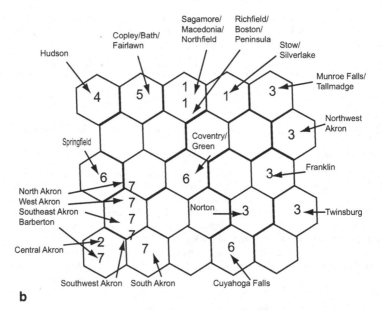

(Cluster Solution for Case Study. The numbers in Figure 2b represent the k-means
cluster solution to which each community belongs. Figure 2b is best read in clockwise
fashion, moving from the most affluent and healthiest communities in the top left, to
the least healthiest communities in the lower left.

Fig. 6.1 Final SOM solution for 20 communities in summit county

grid shown in Fig. 6.1b is called a u-matrix, onto which the SOM clustered our 20 communities—the numbers listed on Fig. 6.1b represent the k-means cluster number for each of the 20 communities. Communities distant from one another in Fig. 6.1b are less alike than those closer to one another. As an addition, Fig. 6.1a is a topographical (3-dimensional) u-matrix. On this u-matrix, gray-scale changes indicate conceptual hills and valleys: the lighter the polygon, the greater the conceptual distance (hill) between cases.

Looking at Fig. 6.1, the cluster solution arrived at by the SOM is very similar to the k-means solution. One can see that the richest communities (Clusters 4 and 5) are close to each other, with cluster 4 located in the farthest upper left corner—this is Hudson, the richest community. Moving along in clockwise fashion, one finds clusters 1 and then 3 (the next most affluent communities) followed by clusters 6 and 7, and then finally Cluster 2. All of the poor communities in Akron are in the lower left corner of the u-matrix, with the poorest community (Central Akron) residing in the farthest lower left corner. (As a side note: later, when we look at how the 17 variables in our study are distributed across the u-matrix—Fig. 6.2 below—the position of the 20 communities relative to one another will make more sense.) The only real difference between the SOM and the k-means is Cluster 6, which the SOM distributed more widely than the k-means. Still, overall, the SOM seems to corroborate the k-means solution, as well as fit with our general, expert knowledge of Summit County.

Goal 2: Our second goal for testing $C2$ was to see what sort of causal model emerged from our cluster analysis. To do so, we used the results found in Fig. 6.2. One of the major strengths of the SOM is that it can project all 17 variables onto the u-matrix, showing how the distribution of each variable helped to cluster and place the 20 communities in Summit County relative to one another. Figure 6.2 is the visual product of the SOM's variable placement. Each of the 17 small maps in Fig. 6.2 represents the distribution of each variable on the u-matrix. For example, the map for Household Income shows that the highest household incomes are in the top left-hand corner and, moving along in clockwise fashion, the lowest household incomes are in the bottom left-hand corner. The location of the lowest household incomes also happens to be the place on the u-matrix where the highest rates of poverty, public assistance, years of life lost and so forth are located.

With the information from Fig. 6.2 in hand, we were able to generate thick causal descriptions for each of the clusters in Fig. 6.1. Consider, for example, Cluster 2, which contained the poorest community in Summit County, Central Akron. Moving from left to right across Fig. 6.2, one sees that, in the 1990s this community (which the SOM locates in the bottom left-hand corner of the u-matrix) had the lowest percentage of whites and the highest percentage of African-Americans; the highest poverty rate; the lowest household income; one of the lowest job growth rates; the lowest workforce percentages; the highest rate of public assistance; the worst graduation record; the highest percentage of unemployment and some of the worst health outcomes indicators, including the worst mortality rate. Finally, it had the greatest number of households where the mortgage/rent was greater than 30 % of income. In

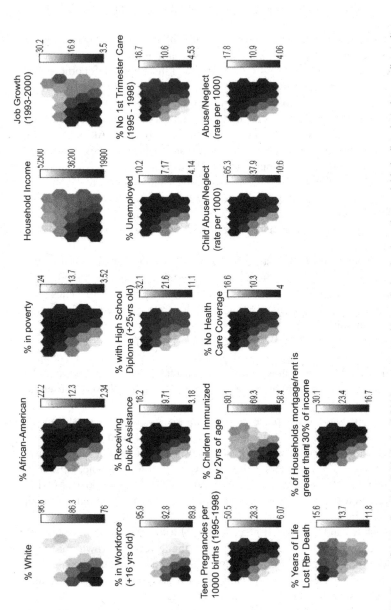

NOTE: For each factor listed in Table 3, the SOM has generated a map, showing how that variable is distributed according to its impact on the U-Matrix in Figure 2. Like the U-matrix, the darker the polygon, the lower the value; the ligher the polygon, the higher the value. For example, the highest household income values are in the upper left of the **Household Income** map--which is where the SOM placed the most affluent communities on the U-matrix in Figure 2.

Fig. 6.2 SOM maps of the 17 variables used to Cluster the 20 communities in case study

fact, when combined with the k-means information in Table 6.1, one begins to develop a rather sophisticated narrative of Central Akron: its median household income is roughly eleven 1000 \$; its job growth between 1993 and 2000 was 20 %; and yet, only 86 % of its labor force was working; and a quarter of the population was on some type of public assistance and did not have health insurance. Furthermore, the average person living in Central Akron lost 16.4 years of life, compared to Hudson, the most affluent community, which (at years of life lost = 10.5), lived an average of 6 more years. Other health outcomes were equally severe. In comparison to Hudson, which has a teen pregnancy rate of 1.3 per 1000; the rate for Akron is 66 per 1000. Less than 40 % of children in Central Akron have received their age-appropriate immunizations by age 2; and, in comparison to the child and elder abuse rates in Hudson, which are almost nonexistent; the respective rates in Central Akron are 98.3 and 53.8 per 1000.

What is even more analytically interesting, however, is when the profile of Central Akron is compared to the communities in Cluster 7. While Central Akron has some of the worst compositional, contextual and health outcomes, Southwest Akron, for example, despite being included in Cluster 7, has a slightly higher mortality rate—with years of life lost per death at 17.4, compared to Central Akron's 16.4. And yet, Southwest Akron was placed in Cluster 7 because, overall, it is doing better in terms of contextual and compositional factors. For example, Southwest Akron's household income level is roughly 7500 \$ higher than Central Akron. Why, then, does Southwest Akron have a higher mortality rate? The answer is not found in any one variable. Instead, as we will explain in our test of $C4$, it is found by looking at the configuration of this community as a whole, over time. Between 1990 and 2000, the socioeconomic health and wellbeing of Southwest Akron spiraled downward— despite job and household income growth. In other words, Southwest Akron is in systems failure.

Of course, if the current article was only about $C2$ or creating narratives about the configurations of these 20 communities, we could go on to construct an increasingly detailed causal model of Summit County, comparing and contrasting configurations; and we could go on to collect more detailed qualitative information to develop our narratives even further. Suffice to say for now that our results validate the main points of $C2$. Furthermore, in terms of a negative test, our results suggest that (a) $C2$ is a theoretically valuable improvement over conventional theory and method, which would struggle to arrive at such a case-based, qualitatively-subtle, variable-complex portrait; (b) $C2$ is not a repetition of what is already known; and (c) because the insights of $C2$ are narrative and visual in form, they are easy to understand.

Chapter 7
Places Are Nodes within Larger Networks

The main point of $C3$ is that the health and wellbeing of places is, in part, a function of the larger socio-spatial networks in which they are situated. As Cummins et al [21] state, "places may be more usefully viewed as nodes in networks than as discrete and autonomous bounded spatial units" (p. 1827).

As we will discuss later, studying places as "nodes in networks" links $C3$ to Characteristic 7 ($C7$): the idea that communities are open-ended with fuzzy boundaries. However, while the main focus of $C7$ is exploring "places as nodes in local, regional and transnational flows of information and other resources" (Cummins et al. 2007, p. 1832); the main focus of $C3$ is exploring "the position of places relative to each other" ([21], p. 1832). As Cummins et al. [21] state: "Studies often ignore issues of spatial autocorrelation [clustering] and assume that conditions in each locality operate on population health independently of conditions in other areas" (p. 1832). In other words, studying places as 'nodes within networks' or exploring the 'autocorrelation amongst a set of geographically proximate communities' is really about mapping and researching communities based on their relative geographical and socio-economic position to one another.

Like $C1$, $C3$ has received empirical attention. In fact, two of the articles in Dunn and Cummins's special edition address this topic: (1) Cox, Boyle, Davey, Feng and Morris [20] and (2) Sridharan, Tunstall, Lawder and Mitchell [65]. In terms of sprawl, $C3$ hold much promise because, as we explained in our introduction, sprawl is, by definition, a network phenomenon that takes place within, between and across an evolving set of communities.

To test $C3$, we used Pajek (see methods section) to explore how communities can be conceptualized as nodes in networks and to see what additional information this conceptualization provided. To do so, we constructed a network map of Summit County and its 20 communities for 1990, using the results we obtained from our cluster analysis in $C2$. (As a side note, our test of $C3$ using 2000 data proved to be equally valid.)

We used the results from our cluster analysis for four reasons. First, if the purpose of our network analysis is to examine the configurations of communities relative to one another, we need (as shown in Table 6.1) the results from our cluster analysis. Second, the u-matrix from our cluster analysis provided us a spatial (albeit non-network)

© Springer International Publishing Switzerland 2015
B. Castellani et al., *Place and Health as Complex Systems,*
SpringerBriefs in Public Health, DOI 10.1007/978-3-319-09734-3_7

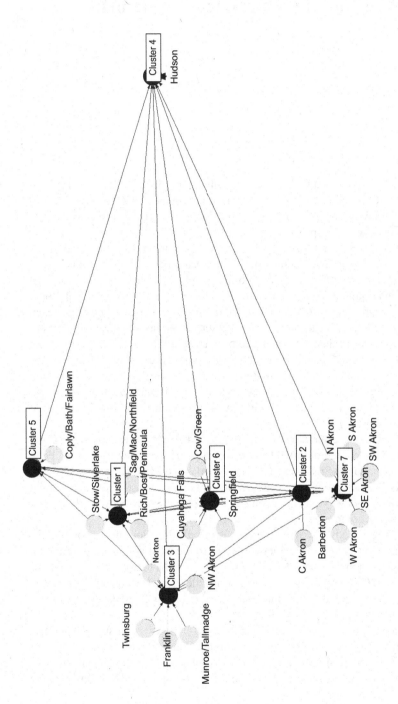

NOTE: Distances between clusters are based on Euclidian distances arrived at through k-means analysis. Distances within clusters for each community are based on within-cluster measures. All measures are non-standardized.

Fig. 7.1 Network map of the seven clusters in summit county and their respective communities

representation of our 20 communities relative to one another, based on their different configurations. Third, our k-means cluster analysis provided within-cluster distance measures for each of our seven clusters. These within-cluster distances are useful because they provide a weighted, spatial representation of how the cases for each cluster auto-correlate based on their position relative to each cluster's center. Finally, our k-means cluster analysis provided us with the un-standardized Euclidean distances between our seven cluster centers. The shortest distance between two points is a straight line. In cluster analysis, a Euclidean distance measure, albeit unstandardized, tells us the shortest conceptual distance between the seven cluster centers in Summit County. While, to our knowledge, nobody has used distance measures to construct a network map of communities, such data are ripe for network analysis as they are weighted, spatial representations of the relationships amongst a set of cluster centers—which are, in turn, spatial representations of the different complex configurations that exist amongst similar communities.

Figure 7.1 is a network representation of our cluster analysis data. The network is made up of our seven cluster centers, labeled 1 through 7. Around each cluster are the communities associated with it. Like Fig. 6.1, the greater the distance between cluster centers, the less alike these clusters are; and, the greater the distance a community is from its cluster center, the less similar its configuration is to the other communities in its cluster.

Looking at Fig. 7.1, the socio-spatial positioning of the seven major clusters and their respective communities provides useful information. First, it supports Sridharan, Tunstall, Lawder and Mitchell's [65] idea that socio-spatial auto-correlation is important. The communities in five of the seven clusters are tightly grouped around their centers, based on their key social measures, suggesting that the communities of Summit County are neither isolated nor alone in the economic struggles or relative health challenges with which they are dealing. For example, Pajek positioned the seven poorest communities (clusters 2 and 7) next to each other. In terms of spatial auto-correlation, all seven communities are also geographically proximate to, as well as socioeconomically interdependent with each other, making up the two interconnected urban centers in Summit County (Akron and Baberton)—see Map 1. The poverty of all seven communities is also socially and spatially interlinked, as Akron and Barberton struggle with issues of out-migration, the collapse of their industrial-based economies, the growing poverty of their citizens, and the failures of their public institutions, particularly education, to assuage these challenges.

The socio-spatial position of Cluster 4 (Hudson) further corroborates $C3$, as it shows how, in terms of sprawl, the spatial wellbeing of some clusters depends upon the deprivation of others. In fact, as will be discussed in our tests of Characteristics 4, 7 and 8, the significant health and wellbeing of Hudson is linked, in part, to the economic struggles of Akron and Barberton and their surrounding communities, along with county-level residential mobility issues and suburban sprawl. Figure 7.1 also corroborates the findings of $C2$, particularly the spatial arrangement of the communities shown in Fig. 6.2, which was generated by the SOM. As such, Fig. 7.1 does a good job visually demonstrating just how qualitatively different Cluster 4 is from the rest of the communities. Finally, the position of Clusters 6, 3, 1 and 5 respectively is in accordance with their increasing health and wellbeing.

Chapter 8
Places Are Dynamic and Evolving

The main point of C4 is that places are best studied as dynamic and evolving. As Cummins et al. [21] state, "[I]t may be just as important for contextual studies to begin to understand not just the life course of individuals, but also the social and economic trajectories of the places which they inhabit" (p. 1832). Or, as Gatrell [30] states, "Complex systems have a history; their past is 'co-responsible' for their present behavior" ([30], p. 2662). However, while the basic point of C4 is clear enough, the methodological challenge is figuring out how exactly to model these ideas; particularly given the somewhat vague and historical manner in which the COP literature uses these terms, in comparison to the more specific and often times precisely mathematical (albeit computational) meaning.

In complexity science, to say that a place is dynamic and evolving means, at its most basic, several key things [42]. First, it means that a place is in a constant state of motion. Second, this motion emerges out of the interactions of its multiple parts (i.e., agents, forces, etc). Third, a place evolves across time/space and does so along a certain set of trajectories—which represent the set of all 'empirically possible' states a place can take at a particular moment, partitioned from the larger set of all 'theoretically possible' states (the state space). Fourth, while dynamic, the evolution of most places falls into a relatively stable pattern, known as its set of attractor points. Sometimes, however, a place's attractor point(s) can turn chaotic or strange—think, for example, of a sudden economic collapse or moments when political unrest in a town or city becomes chaotic, or the sudden exodus of people, as in the case of sprawl. In such instances, the future state of a place is difficult to predict. But, generally speaking, the dynamics of most places, while stochastic, do evolve along a reasonably stable set of trajectories (attractor points).

The goal of our test was to examine the discrete evolution of the 20 communities in Summit County by modeling their state space in 1990 in order to locate the dominant attractor points in the system, and then to compare these results to a second point in time, 2000, to see how this network had evolved and, more specifically, if any of the trajectories had changed; and, if so, how. We accomplished this goal as follows:

1. First, the dataset for the cluster analysis needed to be treated as if it were S. In our study, for example, we treated all 20 communities as if they were part of the

© Springer International Publishing Switzerland 2015   51
B. Castellani et al., *Place and Health as Complex Systems*,
SpringerBriefs in Public Health, DOI 10.1007/978-3-319-09734-3_8

complex system called Summit County. This makes the search for attractor points (trajectories) case-based.

2. Second, k-means and the SOM needed to be employed to identify, map and analyze the most common vector configurations in S for a particular moment in time/space.

3. Third, the obtained cluster centers (centroids/neurons) needed to be treated as S's attractor points. In our study, for example, the seven clusters discussed in our results section became the attractor points in our system (Summit County), around which the 20 communities in our study grouped.

4. Next, the un-standardized Euclidean distance measures provided in the proximity matrix for S need to be mapped to visually identify S's attractor points and the cases clustering around them. In our study, for example, Fig. 7.1 becomes our map of Summit County and its attractor points for 1990. We generated this map by entering our unstandardized weights into Pajek, a network analysis software package. The map was generated using the popular network visualization algorithm, Kamada-Kawai [38].

5. Next, the within-cluster distance measures for the cases needed to be used to map the relative distance of each case to the particular cluster/ trajectory/ attractor point to which it belongs. This is useful because the cases clustering around the system's attractor points become data for creating a thick description of the various trajectories toward which the system is drawn.

   In our study, for example, the communities clustering around each solution were used to construct thick descriptions of the different directions Summit County is heading. Here is a very brief description of what Fig. 7.1 suggests: Summit County seems to have evolved into three, main trajectories: there is a trajectory toward affluence and health, represented by Cluster 4; a trajectory toward poverty and poor health, represented by clusters 2 and 7; and a sort of middle ground trajectory, perhaps representing the major settling point for this County, which revolves around average to above average health and wellbeing, represented by Clusters 1, 3, 5 and 6. Also, although not shown in Fig. 7.1, to further test our notion that Clusters 1, 3, 5 and 6 constitute the major settling point for this County, we re-ran our cluster analysis, entering Summit County as a 21st community, using for its vector configuration the County-level averages for all 17 variables. Our cluster analysis grouped Summit County with Cluster 6.

6. Finally, if such information is available, the above five steps can be repeated for additional points in time. These additional points in time can be then compared to the first point in time to see how the attractor points in the system might change.

For example, for our study we re-ran the above analyses for 2000, to explore how the seven trajectories in Summit County changed over a 10 year period. We used the same entry order as shown in Fig. 7.1, using 2000 data for all compositional and contextual factors. Our health outcomes, however, did not change, as these were aggregated across varying periods of time between 1990 and 2000. For example, Years of Life Lost per Death came from data aggregated between 1990 and 1998. Running our k-means we also sought the same 7-cluster solution found in 1990 to see if it continued to prove useful. Table 8.1 is a quick summary of our results. In terms of

Table 8.1 Change in final cluster solutions for 20 communities in summit county, 1990 to 2000

Community	Year	
	1990 cluster membership	2000 cluster membership
(Affluent cluster) hudson	4	4
(Affluent cluster) Copley/Bath/Fairlawn	5	5
(Middle Class Cluster) Stow/Silverlake	1	1
Northfield/Macedonia/Sagamore	1	1
Richfield/Peninsula	1	5[a]
Twinsburg	3	1[a]
Northwest akron	3	3
Munroe Falls/Tallmadge	3	3
Norton	3	6
Franklin	3	3
Springfield	6	6
Coventry/Green	6	**3**
Cuyahoga Falls	6	6
(Poor Cluster) North Akron	7	7
West akron	7	7
South akron	7	7
Southwest akron	7	**2**
Southeast akron	7	7
Barberton city	7	7
(Poorest Cluster) central akron	2	2

[a]The values listed in the columns for all 7 clusters represent the average value/measurement

reading Table 8.1, the second column shows cluster membership for each of our 20 communities in 1990; the third column shows their 2000 membership. Also, going from the top of the table to the bottom, we ordered the clusters from the most affluent to the poorest.

The first thing that stands out in Table 8.1 is that the seven major trajectories in Summit County continued to exist in 2000. However, there were some interesting shifts. For example, the poorest trajectory, represented by Cluster 2, gained a community, Southwest Akron. On the opposite side, the second most affluent trajectory, represented by Cluster 5, also gained a community. These two shifts suggest that the richest and poorest trajectories not only gained in strength but also, between 1990 and 2000, a widening socioeconomic gap emerged between the poorest and richest communities. This gap was further corroborated when we re-ran our 2000 k-means with Summit County included. The County dropped from Cluster 6 to Cluster 7, which contains most of the poor communities and, after Cluster 2, the worst health outcomes.

There is a lot more to be explored in our analysis of the evolution of the network of communities in Summit County. In fact, as discussed in our review of the SACS Toolkit, we devoted an entire study to modeling the continuous dynamics of Summit County by employing our case-based density approach, which makes use of genetic algorithms, ordinary differential equations and the advection (partial differential equation) equation(See Rajaram and Castellani 2013). Our 2013 study likewise confirmed the importance of conceptualizing and modeling communities as dynamic and evolving. However, given the need to report on our second test, we must stop here, content with the fact that the test accomplished its goal: it determined that the evolution of a network of communities could be effectively modeled with the SACS Toolkit and that such an analysis does provide theoretically novel insights into the evolution of this complex system.

Chapter 9
Places Are Nonlinear

Scholars involved in the study of place are very clear about how they define and intend the characteristics of nonlinearity. As Gatrell [30] and others explain (e.g., [24, 25], [28]), in terms of a complex system like place, nonlinearity addresses the empirical fact that, more often than not, small or large changes in some aspect of a place (e.g., its health system, educational system, etc), particularly in the form of health interventions (e.g., new outpatient program, new educational accountability measures, etc) do not regularly lead to their expected, linearly related outcomes (e.g., a 25 % increase in prenatal care or graduation rates, etc.).

Our test of $C5$ attempted to examine if it is realistic to assume that, given the complexities of sprawl, and its unequal impact on community-level health, that health interventions by a County to improve its communities are, indeed, nonlinear in their outcomes. For our test, we focused on a report released by the Summit County Social Services Advisory Board (SSAB), titled Summit 2010 Priority Indicators Progress Report, 2009 (http://www.healthysummit.org/Summit2020QoL.html).

In 2003, the SSAB began a program it called Healthy Summit 2010, a community-level version of the federal government's Healthy People 2010. The purpose of Summit 2010 was to establish a set of 20 goals for improving the economic, institutional, physical and behavioral health and wellbeing of Summit County. In turn, as shown in Table 9.1, these 20 goals were turned into priority indicators that the SSAB could use to measure the County's progress.

In terms of testing the concept of nonlinearity, two aspects of the Healthy Summit 2010 project need to be addressed: effort and outcomes. In terms of effort, it is clear from reading the Healthy Summit 2010 Quality of Life Project website that the work being done in Summit County to improve the health and wellbeing of its citizens, particularly those in need, is ambitious both in scope and in effort. This ambitious, concerted effort includes the SSAB; its numerous committees; directors and researchers; an extensive network of health and social service systems; and a long list of health care providers, politicians, community leaders and activists. In fact, as stated on the front page of the website: "The Quality of Life Project, under the guidance of the Social Services Advisory Board, will lead Summit County to new, unprecedented success in the age-old battle to improve health, expand economic opportunity, and reduce poverty and its ill effects."

© Springer International Publishing Switzerland 2015
B. Castellani et al., *Place and Health as Complex Systems,*
SpringerBriefs in Public Health, DOI 10.1007/978-3-319-09734-3_9

Table 9.1 Health outcome indicators for healthy summit 2010

Priority indicator	Description	Indicator currently better, same, or worse than in 2003 environmental scan
Indicator 1	Increase the proportion of people living above the official poverty line	Worse
Indicator 2	Increase the proportion of African-Americans living above the poverty line	Worse
Indicator 3	Reduce unemployment	Worse
Indicator 4	Increase the proportion of people aged 25 and over who have received a high school diploma	Better
Indicator 5	Increase housing affordability, raising the proportion of households spending under 30 % of their incomes on housing	Worse
Indicator 6	Reduce the proportion of households receiving Temporary Assistance for Needy Families (TANF)	Better[a]
Indicator 7	Reduce the incidence of domestic violence-related crime	Worse
Indicator 8	Reduce the rate of violent crime	Comparable data currently unavailable
Indicator 9	Increase the proportion of African-American children under age 5 living above the official poverty line	Same
Indicator 10	Increase the proportion of children receiving their immunizations by their second birthdays	Worse
Indicator 11	Reduce the incidence of child abuse and neglect	Better
Indicator 12	Increase secondary school attendance	Better
Indicator 13	Increase the proportion of African-American children age 18 or less living above the federal poverty level	Same
Indicator 14	Reduce the rate of births to teens, focusing on higher rates among African-American youth	Better
Indicator 15	Increase the proportion of African-American older adults (age 65+) living above poverty	Same
Indicator 16	Increase self-sufficiency of seniors living alone	Same
Indicator 17	Reduce the incidence of elder abuse and neglect	Same
Indicator 18	Increase the proportion of individuals with health insurance	Worse

Table 9.1 (continued)

Priority indi-cator	Description	Indicator currently better, same, or worse than in 2003 environmental scan
Indicator 19	Increase the proportion of pregnant women receiving first trimester prenatal care	Better
Indicator 20	Reduce the rate of years of potential life lost from all causes	Same

Better-The indicator improved relative to the 2003 Environmental Scan; **Same**-The indicator did not show any appreciable change from the 2003 Environmental Scan; **Worse**-The indicator declined relative to the 2003 Environmental Scan; **Unknown**-Because of missing data or overlapping confidence intervals, it is unknown whether any change in the indicator occurred.

[a]While the goal of reducing the percentage of the population on public assistance has been met, the "improvement" seen is most likely being caused by needy families using up their current eligibility for public assistance rather than an increase in economic self-sufficiency.

Source: This Figure was taken from the Summit County "Priority Indicator Progress Report 2009"

The challenge, however, is that, between the years of 2003 and 2009 there was no statistically significant, linear relationship between the concerted efforts of the Healthy Summit Quality of Life Project and their 20 health outcomes. In fact, looking at Table 9.1, the data show that improvements across the Healthy Summit 2010 twenty key indicators were, at best, uneven. For example, while a few indicators, such as Education (Indicator 4) and Teen Birth Rate (Indicator 13) showed some level of improvement, other indicators, such as General Poverty (Indicator 1) and Unemployment (Indicator 3) got worse; and other indicators, such as Years of Potential Life Lost (Indicator 20) remained the same.

As stated previously, the nonlinear relationship between effort and outcome is a major challenge for community health providers (e.g.,[40]). It is usually unclear, for example, what types of concerted efforts will yield the measurable outcomes desired. For example, while research on chaos theory (e.g., [33]) and tipping points (e.g., [32]) suggests that small changes in initial conditions can lead to sudden and significant change in outcomes; research on poverty traps (e.g.,[8]) and nonlinearity in complex human organizations [51] suggests that significant efforts often lead to little difference. Related, it is not always clear how best to measure change. For example, will a community's efforts to address poverty yield 5 years of no results, only to suddenly produce significant change by year 6? Such questions are extremely important in an era of fiscally-driven policy that often lacks equal theoretical or empirical rigor. The results here suggest that, without a meaningful understanding of the nonlinear dynamics of communities and the interventions into them (particularly by thinking about how communities are parts of larger complex networks) there is a significant chance for money and time and effort to be misused and for people, who could otherwise be more effectively treated, to go without the care they need.

Nonetheless, we must, at this point in our test, stop. While we cannot address these questions or issues in the current study, our quick test of Summit County suggests that, in complex systems such as communities, the relationship between public health interventions and their outcomes can be rather nonlinear.

Chapter 10
Places Are Subjective and Historical

Like $C1$, $C6$ has a positive and critical form. In its positive form, its point is that subjectivity, personal experience and history play an important role in the dynamics of places. This part of $C6$ is self-evident and does not need testing. The critical form of $C6$, however, does need to be addressed. It basically argues that, while scholars involved in the study of place know that history and subjectivity play an important role in health, they ignore this type of research, opting instead for quantitative analysis. As Cummins et al [22] explain: While a call has been made to explore health as the "lived" and "embodied" experiences of people interacting with their settings, these ideas "remain poorly integrated into empirical research" (p. 1829). To address this lack of integration, researchers need to incorporate into their work "information about settings that are drawn from reported views of residents, as well as from independently measured indicators of local conditions" ([21], p. 1830).

While readers may not know, complexity scientists almost unanimously share the same "anti-qualitative" bias as the study of place—by qualitative we mean here a bias away from field studies or narrative or historical data and archives. In fact, following Smith and Jenks [63] and others (e.g., [15, 17]), there is almost no qualitative or historical research done in complexity science. For example, as Gatrell [30] states, "[T]he human voice seems to be missing from much of complexity theory. The qualitative is there, but in the form of qualitative structures and patterns, not in the nature of the embodied actor" (p. 2669). And so, pace our Definitional Test of Complex Systems (DTCS), the inclusion of $C6$ in Table 1.1 constitutes a novel and important addition to the otherwise obstinately 'natural-science' based definition of complex systems typically imported into the social and health sciences.

Given the self-evident nature of $C6$ to (at least) social scientists, the purpose of our test was to see if qualitative information improved our theoretical understanding of Summit County and the impact sprawl is having on it. For our test, we used two qualitative reports located on the Healthy Summit 2010 Quality of Life Project website Visit the website at: http://www.healthysummit.org/Summit2020QoL.html).

First, there is the *Neighborhood Project Summary Report*. This report summarizes a series of focus groups conducted with members living in three targeted neighborhoods in Summit County: Barberton, Buchtel and Lakemore. Second, there is the *Report on Key Informant and Community In-person Interviews*. This report

© Springer International Publishing Switzerland 2015   59
B. Castellani et al., *Place and Health as Complex Systems*,
SpringerBriefs in Public Health, DOI 10.1007/978-3-319-09734-3_10

was based on $N = 230$ interviews: 190 interviews with key informants from the public, nonprofit and private sectors in Summit County; and 40 interviews with county residents—two from each of the 20 communities in Summit County. We used these reports because they are qualitative in nature, giving attention to issues of voice and subjective experience; and because these reports were used to develop the Healthy Summit 2010 Quality of Life Project, including its strategic interventions for improving health.

Reading through these reports, we found them to be rich with qualitative insight into the complexities of place and health in Summit Count, both across the county as a whole and within the poorer communities. We do not have the time to delve into these details here. Nonetheless, we can highlight a handful of key points:

For example, the *Neighborhood Project Summary Report* (NPSR) engaged local citizens in three neighborhoods to establish a set of goals for addressing the major health care problems they face. The NPSR for Barberton (the second largest city in Summit County, which struggles with significant poverty) was particularly interesting, as the topic of health was directly connected to employment, not health policy intervention. The NPSR summary for Barberton states, "The primary concerns of the residents, elected officials, and professionals uniformly 'centered around' employment. Educational concerns, family problems, and crime were all seen as directly related to employment difficulties. The loss of jobs in the city, education not matching employment opportunities, and difficulties with transportation to employment and training opportunities were cited as areas needing intervention" (NPSR, p. 10).

For us, this summary goes to the heart of the importance of qualitative inquiry for understanding the complexities of place and health. Without the need for complex modeling procedures, the citizens of this community were quite aware that sprawl and the out-migration of jobs and resources and money had left their community struggling with significant employment issues—which has directly impacted their community's health and wellbeing. In turn, for these folks, policy intervention needs to focus on sprawl and out-migration and keeping jobs and resources and capital in the community; not just triaging for a problem that is only going to get worse.

The *Report on Key Informant and Community In-person Interviews*—which was based on first-person interviews—also had some important things to say the social factors responsible for sprawl and the out-migration of affluence and resources in Summit County to the suburban communities. These factors included: (a) racism; (b) the negative perception affluent communities have of public services; (c) affluent flight from Akron (the main city in Summit County) to the suburbs; and (d) political and economic turf battles between certain communities, as affluent folks moved out into the suburbs.

It is important to note that none of the above issues were measured as a compositional or contextual variable in the *Healthy Summit 2010* reports, primarily because they constitute the nuanced complexities of context and the manner in which macroscopic social problems (in the form of social structure) flow through, around, in and across the communities in this county—which is exactly why the COP approach includes it in its 9-characteristic definition. And, it is exactly the same reason why the authors of *Healthy Summit 2010* included this report in their study: qualitative information plays an important role in understanding the dynamical complexity of places and their health.

And, this is not where the richness of this second report ended. It went on to solicit opinions from people on the strengths and weaknesses, opportunities and threats for a host of health issues in some of the poorer communities, including (a) effectiveness of community-based organizations; (b) children and youth programs; (c) services for working adults; (d) family services and (e) elderly care. While we cannot delve into these opinions, suffice to say we found them rich with information about the historical, political, economic, cultural and institutional nuances of Summit County and its health and health care, allowing us to conclude that this type of information is a valid and valuable part of understanding the structure and dynamics of communities as complex systems.

Chapter 11
Places Are Open-Ended with Fuzzy Boundaries

The main point of $C7$ is that places are not autonomous entities or closed-systems with clear geographical or sociological boundaries. Instead, they are open systems with fuzzy boundaries. Like many of the characteristics in Table 1.1, $C7$ is well defined. Let us explain.

In complexity science, an open system is one that interacts with its environment. Almost all biological life forms are, in principle, open systems insomuch as they survive through their adaptive interactions with the environmental mediums in which they live ([16, 48]). Places are similar. However, in the case of contemporary, globalized society, survival not only depends on a place's interactions with the physical environment, but also the larger globalized, economic, political and cultural systems in which it is situated.

The open-ended nature of places takes us back to $C3$ and the study of social networks. As a reminder, while $C3$ focused on communities as "nodes within networks," $C7$ focuses on communities "as nodes in local, regional and transnational flows of information and other resources" (Cummins et al. 2007, p. 1832). For example, Cummins et al. state, "We can conceive, for example, that trends [flows] in regional economies, national and regional environmental pollution, national or supra-national organizations and entities can all define the 'local' and other contexts in differing ways and this in turn contributes to the spatial distribution of health outcomes" (2007, p. 1833). For some communities, such flows can produce significant improvement in overall wellbeing, as affluence and resources move into them; in turn, for other communities, such flows can lead to downward wellbeing and, in the worst case scenario, poverty traps.

In addition to being open-ended, the boundaries of communities tend to be fuzzy. Fuzziness has to do with mobility—perhaps the most important factor that engines sprawl. In a global society, the agents in a community (be they people, businesses or health care systems) often do not confine themselves to the geographical or sociological boundaries of their home communities [67]. Health care systems, for example, tend to spread out, building their centers in other communities or caring for patients within a particular region. In turn, in sprawling counties like Summit County, people often live in one community while working in another; get their groceries in one place while receiving their health care in another. The social networks of people also

© Springer International Publishing Switzerland 2015
B. Castellani et al., *Place and Health as Complex Systems,*
SpringerBriefs in Public Health, DOI 10.1007/978-3-319-09734-3_11

bleed across communities, especially when one allows for the influence of cyber-infrastructure (e.g., cell phones and online social networks). We could go on. The point is that mobility is a major issue in terms of defining the boundaries of places, so much so that it is best to: (a) view place boundaries as fuzzy and (b) define the "place-level" exposure of many individuals as multiply determined. As Cummins et al. [21] state, "These issues of varying individual-level exposure to multiple contexts over time and space means that current measures of simple universally applied 'neighborhood' exposure may severely underestimate the total effect of 'context'" (p. 1830). Again, this 'context' could be in the form of a highly mobile community or in the form of highly stagnant communities, stuck in poverty and health traps, both socially and spatially.

To test $C7$ we turned to a study recently completed by the *Northeast Ohio Urban Sprawl Modeling Project* (NOUSMP)—See Peterson et al. [51]. The purpose of NOUSMP was to provide the general public a data-driven visualization of the projected impact that sprawl will have on the 15 counties of northeast Ohio by 2020, including Summit County (See pp. 2–4).

Figure 11.1 was created from maps generated by *ArcExplorer*, a public domain software program for viewing and studying geographical information systems (GIS) data. The GIS data for Fig. 11.1 came from a variety of environmental and geospatial databases (pp. 2–14), which NOUSMP researchers entered into their model. The projected spatial distribution of Summit County's population, as shown in Fig. 11.1 Map B, was based on a simple set of assumptions about how urban sprawl would occur between 1990 and 2020, such as projected residential growth along highways (pp. 2–4). Putting all of this together, Fig. 11.1 shows the change in population density in Summit County as a function of urban sprawl in northeast Ohio (circa 1990, Fig. 11.1, map a) and its projected impact in 2020 (Fig. 11.1, map b).

Here is how Fig. 11.1 helps us test $C7$. Maps A and B in Fig. 11.1 are organized according to the major communities of Summit County. The lines moving out of Akron denote the major highways and roads in Summit County. In terms of population density, the darker the color, the denser the population. Looking at Map A, the highest rates of population density in Summit County are found in the communities surrounding the city of Akron, with almost no major density along any of the highways. Map B shows a different story. By 2020, the population density of Summit County shifts away from the cities, concentrating around most of the major highways and roads of the second-tier suburban communities of Summit County—see the arrows in Map B.

The shift in population density shown in Fig. 11.1 is a small window into how the boundaries of the communities of Summit County have become increasingly open-ended and fuzzy. In terms of their open-ended nature, sprawl has forced the communities of Summit County (as nodes within a larger network) to adapt to the wider regional flows of commerce, people and capital, as well as the impact these sprawling flows have on the 'local' context. And, in terms of fuzziness, as people and jobs move into Summit County's suburban communities, they leave behind the poor. These insights happen to corroborate the 'first-person' narratives we discussed in $C6$, where outmigration and the movement of jobs elsewhere are seen as a major

MAP A: 1990 Population Density   MAP B: 2020 Population Density

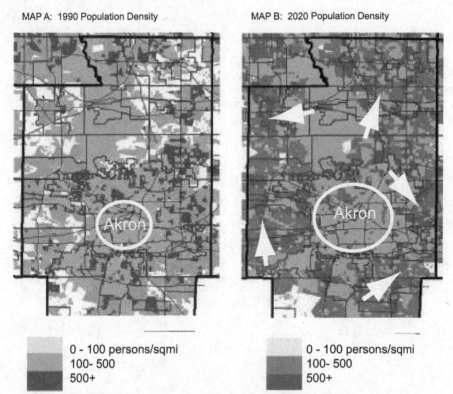

	0 - 100 persons/sqmi			0 - 100 persons/sqmi
	100- 500			100- 500
	500+			500+

The Arrows in Map B show how the population in Summit County shifted away from Akron, moving outward along the major highways and roads found in the second-tier suburbs. See Map 1 as a point of comparison

Source: This map was retrieved from http://gis.kent.edu/gis/empact/index.htm on the 29th of November, 2011. It is a public, USA government document (US EPA Grant #985989-01-0).

Fig. 11.1 Maps of summit county population density (1990–2020) as a function of Urban Sprawl

barrier to improving the economic wellbeing of poor communities in Akron. We could go on, but we need to stop. Our brief review of sprawl does suggest, however, that thinking about places as open-ended and fuzzy is valid and valuable.

Chapter 12
Places Are Power-Based Conflicted Negotiations

$C8$ constitutes another addition made by community health science scholars to the quasi-generalist definition of complex systems used in complexity science. It is clear in its focus: places are imbued with power relations. Its empirical study, however, and its epistemological-theoretical frame need to be developed—particularly given the availability of different theories of power, from Marx to Foucault. Nonetheless, despite its tentative outline, $C8$ makes an important point that complexity scholars involved in the study of place need to address. Let us explain.

For a sociologist to state that places are imbued with power relations is not, at first glance, saying anything new. In fact, only the opposite sort of statement, the idea that places are not imbued with power relations, would require empirical test. Interestingly enough, what is obvious to one field of study or discipline is not necessarily obvious to another. Such is the case in complexity science. For all their 'sociological' discussions about the complexity of businesses, governments, financial markets, social networks, small group dynamics and communities, the majority of scholars in complexity science—particularly those coming from the natural and artificial sciences—have not developed a vocabulary for discussing the impact that power relations (particularly inequality, oppression or exploitation) have on the structure and dynamics of complex systems. And, there is even less discussion of related power issues such as ethnicity, social class or gender. For example, as Gatrell states, "Gender too seems to be a missing strand from existing uses of CT [complexity theory]" ([30], p. 2669). The only caveats to this dominant trend are a handful of scholars working at the intersection of sociology and complexity science (see [17]) and, in terms of the current study, scholars working at the intersection of complexity science and community health. At second glance, then, $C8$ is less a reminder to social scientists and more a challenge to complexity scientists.

To test of $C8$ we focused on a conflict that expressed well how sprawl, as a complex systems problem, results in unintended health inequalities at the community level. The conflict has to do with the two major health systems in Summit County, the Summa Health System (SUMMA) and Akron General Health System (AGHS). The conflict (which began circa 2008) centered on SUMMA's desire to build a 100-bed, physician-owned hospital near Hudson, the most affluent community in Summit County. Historically, the majority of facilities run by SUMMA and AGHS have

© Springer International Publishing Switzerland 2015
B. Castellani et al., *Place and Health as Complex Systems*,
SpringerBriefs in Public Health, DOI 10.1007/978-3-319-09734-3_12

resided within and around the city of Akron, which has been to the health advantage of Akron residents, particularly the poor. The building of a new hospital near Hudson and the affluent communities around it therefore signaled a worrisome change to the communities in Akron.

In response, on October of 2008, AGHS held a forum entitled, "Should Doctors Own Hospitals? And Why Should You Care? A Close Look at Risks and Realities" [46]. In their forum, AGHS explained that its main concern was that SUMMA's new facility, given its location and for-profit status, would pull physicians from the Akron area; physicians would also be able to "cherry pick" their patients, and those affluent patients out-migrating to or presently living in Hudson and the surrounding affluent suburban communities would stop going to Akron to receive care – all of which would negatively impact community hospitals and clinics in Akron, primarily in the form of "huge drops in volumes and revenue," "lowered bond ratings," and "staff layoffs and reductions," which would lead to poorer health outcomes [46]. SUMMA countered, arguing that no such thing would happen; instead, the new facility would increase access for residents migrating to or currently living outside of Akron, as well as stimulate job growth for Summit County. In terms of the current study, as might be expected, the political and economic leaders of the suburban communities vying for the location of SUMMA's new hospital saw the project in entirely positive terms; while the communities that perceived the project as problematic—those within clusters 2 and 7 of our research—saw it as corrosive to the health and wellbeing of Akron communities and, on the whole, to Summit County. SUMMA, through its partnership with the Western Reserve Hospital Partners, went on to build its hospital.

While we cannot go into greater detail about the conflict between SUMMA and AGHS and their respective communities, it illustrates the main point of $C8$, as it relates to the issue of sprawl: places and their health are imbued with power relations that can have a significant impact of the health of their respective residents. We turn, now, to our final characteristic.

Chapter 13
Places Are Agent-Based

Like $C2$, $C9$ draws on (imports) one of the most important characteristics (areas of study) in complexity science: the idea that complex systems emerge out of the self-organizing, adaptive interactions amongst a set of rule-following agents.

To say that places are agent-based means five things to community health scientists. First, it means that interacting agents play a major role in the self-organization, emergence, structure and dynamics of places and their health.

Second, it means that "places are produced and maintained by the activities of 'actors,' proximate or distal to a particular place, who operate individually or in concert across a wide range of geographical scales" ([21], p. 1828). This is a particularly useful point in terms of sprawl, which concerns the impact micro-level actions at a distance (such as moving to the suburbs) has on community health in another location.

Third, it means that it is impossible to think about compositional and contextual factors without considering how they are enacted through the complex interactions of the agents involved in a place. Again, another key point for studying sprawl and health.

Fourth, and more specifically, it means that "place effects on health emerge from complex interdependent processes in which individuals interact with each other and their environment and in which both individuals and environments adapt and change over time" ([2], p. 1). In terms of sprawl, this is very important because it helps us understand the role agency plays in the evolution we saw in our test of $C4$.

Finally, it means that agents "can be conceived of in a variety of ways from individuals and community organizations, firms and businesses, regional and national governments and institutions, peer-networks and families to static and dynamic regulatory structures and processes such as national tax policy and the rule of law" ([21], p. 1828).

In its critical form, the main point of $C9$ (like $C1$ and $C2$) is that multi-level analysis (regression) is an insufficient method for studying the impact that compositional and contextual factors have on health because it cannot account for the role that heterogeneity, agency and interaction play in the structure and dynamics of communities. As Auchincloss and Diez Roux state, "In general, regression approaches continue to be ill equipped to investigate the processes embedded in complex systems

© Springer International Publishing Switzerland 2015
B. Castellani et al., *Place and Health as Complex Systems*,
SpringerBriefs in Public Health, DOI 10.1007/978-3-319-09734-3_13

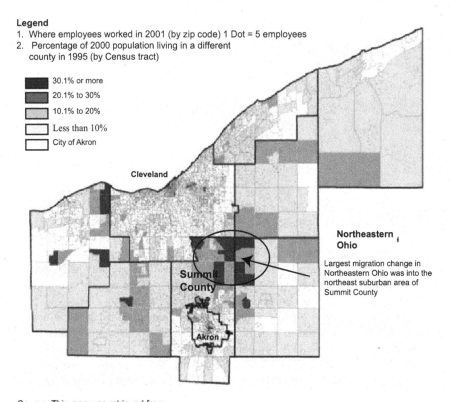

Legend
1. Where employees worked in 2001 (by zip code) 1 Dot = 5 employees
2. Percentage of 2000 population living in a different
    county in 1995 (by Census tract)

    30.1% or more
    20.1% to 30%
    10.1% to 20%
    Less than 10%
    City of Akron

Cleveland

Northeastern
Ohio

Largest migration change in
Northeastern Ohio was into the
northeast suburban area of
Summit County

Summit
County

Akron

Source: This map was retrieved from
http://www.healthysummit.org/qol/pdf/final draft 2009 data tracking report 0216.pdf
on the 29th of November, 2011. It is a public document provided by the Health Summit 2010 website. Its
original source is ES-202 (ODJFS); NODIS, U.S. Census Bureau, 2000.

Fig. 13.1 Change in residential mobility between 1995 and 2001

characterized by dynamic interactions between heterogeneous individuals and interactions between individuals and their environments with multiple feedback loops
and adaptation" ([2], p. 2). As a result, the limitations of multi-level regression "have
constrained the types of questions asked [by community health scientists], the an-
swers received, and the hypotheses and theoretical explanations that are developed"
([2], p. 1). Agent-based modeling, however, acting as a complementary method, can
help to overcome these limitations.

To test the validity of C9, we created an agent-based model called Summit-Sim—
see Method Section for detailed overview of how we built this model. The goal of
Summit-Sim was to test a specific aspect of sprawl: residential migration patterns.

In addition to the report reviewed in our test of C7, the empirical basis for how we
built Summit-Sim came from the Summit 2010 Priority Indicators Progress Report,
2009. In this report, sprawl is used to explain differences in community-level poverty
rates across Summit County. To understand the link between sprawl and differences
in community-level poverty, we need to review Fig. 13.1, which was included in the
report.

Figure 13.1 is read as follows. Dots represent the spatial location of jobs in Summit County and its larger region, Northeastern Ohio, which includes the City of Cleveland and a total of eight counties, including Summit. Each dot represents five jobs. Shaded areas on the map correspond to changes in residential density across the communities in Summit County and Northeastern Ohio. To compute residential density change, researchers looked at where people were living in 2000 compared to where they were living in 1995, arriving at a percentage of the 2000 population in a community that is new. Looking at Fig. 13.1, the northeastern suburbs of Summit County appear to be the major "hot spots" for residential migration within the Northeastern Ohio region.

Figure 13.1 corroborates or helps explain several of the findings we discussed in several of our previous tests. First, while Fig. 13.1 provides a different picture than Fig. 11.1 (C7), it supports the simulated and projected findings of the Northeast Ohio Urban Sprawl Modeling Project: Summit County is a microcosm of the larger region, as residents are moving out of the urban centers of Northeastern Ohio into the suburbs. Second, Fig. 13.1 helps explain the findings at the end of C4 (as a reminder, the communities in clusters 4 and 5 pulled further away from the rest of the County in 2000; and Cluster 5 gained Richfield, Boston Mills and Peninsula) by suggesting that this spatial inequality may be due to affluent residents moving into these suburban areas. Third, Fig. 13.1 supports the argument made by Akron General Health System in C8 that unplanned growth—specifically the movement of for-profit hospitals and health care into the northeastern suburbs of Summit County—may negatively impact the health of residents living in Akron, as money and services move out of the urban areas and into the suburbs.

Together, then, Fig. 13.1 and the results of C4, C7, and C8 can be used to understand how residential (mobility) migration patterns in Northeastern Ohio impact differences (inequalities) in the health and wellbeing of the communities in Summit County; in particular, they can possibly explain differences between the health of the most affluent, suburban communities (clusters 4 and 5) and the poverty and declining health of the poorest urban communities (clusters 2 and 7).

Stated in formal theoretical terms, the above results on sprawl suggest that the agent-based (mobility) migration behaviors of a set of heterogeneous agents (residents) living in Northeastern Ohio changed the residential composition of Summit County between 1995 and 2000. This change occurred, in large measure, as more affluent agents migrated to more affluent suburban communities, leaving behind their less affluent neighbors. In turn, these residential migration patterns resulted in a spatial segregation of health: poor agents remained confined within or found themselves migrating to poor neighborhoods with poor health outcomes (clusters 2 and 7), while more affluent agents migrated to suburban areas with high health outcomes (clusters 4 and 5). Such a trend has the possibility of being reversed if Summit County addressed the challenges of sprawl, as outlined in C7, and the health consequences of this sprawl, as suggested in C8.

13.1   Enter Agent-Based Modeling

While the above theoretical statement is empirically rigorous, it would be very useful
to somehow experimentally test its fundamental assumptions. But, how does one
do that? One possible way is with agent-based modeling ([29, 31]). Agent-based
modeling is useful to community health science because it can experimentally explore
(in the form of a thought experiment) empirically-derived theories on the relationship
between context, composition and health [2]. More specifically, it is useful because
it is good at modeling how the macro-level patterns of communities (e.g., residential
patterns and community-level health) emerge out of the nonlinear, dynamic, micro-
level behaviors of their interacting and intersecting heterogeneous agents (e.g., sprawl
and residential migration behaviors). To explore how this is done, we turn to our
model.

To test $C9$, we explored two questions. First, do the sprawling migration patterns
of the heterogeneous agents living in Summit-Sim result in the clustering and spatial segregation (distribution) of affluence we see in Summit County? As shown in
Fig. 13.2, the answer is: "Yes, there do seem to be some interesting similarities."

Figure 13.2 is a snapshot of Summit-Sim with a preference rating of 3 for all
agents. This rating means that, for each iteration of Summit-Sim, rich agents sought
to live in a neighborhood with three or more rich agents; middle-class agents also
sought to live in a neighborhood with three rich agents; if they could not migrate
to such a neighborhood, they sought to live near other middle-class agents; if they
found themselves in a neighborhood with four or more other middle-class agents,
they stayed; finally, poor agents sought to live in neighborhoods with three or more
middle-class agents if they could; if they could not, they stayed where they were.

Of the various preference ratings available for our model, we chose 3 because it is
a rather modest preference. What made Schelling's model of segregation so powerful
is that macro-level patterns of significant segregation resulted from very mild preference ratings. Sprawl seems to follow a similar pattern. Mild levels of neighborhood
preference should produce significant spatial clustering and segregation.

A visual inspection of Fig. 13.2—which, given the constraints of time and space,
will suffice for our analysis of Summit-Sim—shows that, as expected, a preference
rating of 3 leads to significant spatial clustering and to even more extreme patterns of
segregation than that found in our empirical analysis of Summit County. There are
very tight clusters of rich agents (see Cluster A), surrounded by a few middle-class
agents; there are large, loose clusters of middle-class agents spread out in the same
basic area, moving from the top-right corner of Summit-Sim to the bottom lower-left
corner. Finally, there are tight clusters of poor agents (see Cluster B), some of which
are very large.

The second question we explored was: If the sprawling migration behaviors of our
heterogeneous agents leads to spatial clustering, does this segregation of affluence
result in community-level health inequalities, as seen in Summit County? As shown in
Fig. 13.2, the answer is: "Again, yes; there do seem to be some interesting similarities
suggesting this to be the case." From the start of the model to its completion, the

NOTE: Rich Agents = Squares; Middle Class Agents = Stars; and Poor Agents = Triangles. **Cluster A** identifies one of the dense clusters of rich agents. **Cluster B** identifies one of the dense clusters of poor agents; which complexity scientists would call a poverty trap.

Fig. 13.2 Snapshot of summitSim with a preference rating of 3 for all Agents. (Rich Agents = *Squares*; Middle Class Agents = *Stars*; and Poor Agents = *Triangles*. **Cluster A** identifies one of the dense clusters of rich agents. **Cluster B** identifies one of the dense clusters of poor agents; which complexity scientists would call a poverty trap

context-dependent unhealthiness of poor agents never dropped below roughly 50 %. Meanwhile, the rich agents had near perfect health. These healthiness ratings fit with our analysis of Summit County; in particular, our comparison of the poorest clusters with the more affluent.

While our abbreviated analysis of Summit-Sim leaves numerous issues unexplored—for example, how do different preference ratings impact spatial segregation or health?—it is adequate to support the point of $C9$: the health and wellbeing of communities seems, to a degree, agent-based, such that studies of composition and context should include some form of agent-based analysis.

Still, while it is useful to think of places and their health as agent-based based, there are two significant limits to this view, which the COP definition needs to address. First, following Byrne and Callaghan (2013), an agent-based approach (while useful and insightful) remains something of a fiction, as it tends to be insufficiently evidence-based or data-driven. In the current study, for example, we constructed several such fictions: (1) we broke Summit County into three simplified groups (poor, middle class and rich); (2) we simplified residential preference to a linear function based on the limitations of a 2-dimensional space; (3) we reduced the capacity to migrate and change residence to a matter of velocity; and (4) we limited health to the number of rich agents in a neighborhood. Given such fictions, one needs to be careful about generalizing our findings too far.

Second, as Byrne and Callaghan also argue (2013), agent-based modeling needs to get beyond its micro-determined conceptualization of emergence, where social structures are only the byproduct of agent-based interactions. Instead, social structures (enduring patterns of social organization) need to be viewed as having causal powers; and, in turn, (2) agency needs to be viewed as something that transcends narrow rules for behavior (p. 51). In the current study, for example, we did not build into our model any rules regarding social structure. We were entirely focused on seeing if we could generate, through microscopic interactions, the macroscopic emergent behavior we found in our empirical study of Summit County—which we did. But, it leaves open the question, how valid are our results, given that the constraints of social structure were not included?

We turn, now, to the conclusion of our study.

Chapter 14
Conclusion

As this study has attempted to show, the *complexities of place* (COP) approach is correct that there is value to be gained from conceptualizing places and their health as complex systems. There is also value in the nine-item definition of complex systems that is currently used, at least as pertains to the topic of sprawl and health. Considering all nine tests together, here are our conclusions regarding the utility of the COP approach and, more specifically, its utility for understanding sprawl and health:

1. In terms of the individual tests, $C1$ demonstrated the limitations of linear modeling for studying sprawl and health. Researchers most likely need to re-think the issue of multicollinearity (the issue of predictor variables being highly correlated). If one takes a case-based profile view of place and health, multicollinearity may not be something simply or only to control for; instead, it might also be something to explore and model, pushing us to rethink what is a causal model.
2. Our test of $C2$ further supports the COP challenge of rethinking causality, given that the variables (factors, actors, forces, etc) in a complex system are self-organizing (form a causal arrangement of their own making) and emergent (forming a causal arrangement where the whole is more than the sum of its parts). In light of such insights, how can researchers continue to study individual variables alone?
3. Our tests for $C3$ and $C7$ take the insights of $C1$ and $C2$ even further. Not only are the k-dimensional profiles of places and their health usefully viewed as case-based, complex, self-organizing and emergent; it also seems that, as cases, these profiles of difference influence each other (spatially and sociologically) through their socio-spatial network connections and through the larger, macroscopic impact of the network, itself, as a self-organizing emergent system.
4. Our test of $C4$ showed strong support for exploring how a sprawling network of places evolves across time/space. Despite significant advances in stochastic methods for modeling cohort and longitudinal data, it remains the case that modeling the temporal and spatial dynamics of complex system is a serious challenge. This is of particular concern, given that places and their health, including the larger complex socio-spatial networks and systems of which they are a part are so incredibly dynamic. Public health scholars need, therefore, to spend time significantly reflecting on $C4$.

© Springer International Publishing Switzerland 2015
B. Castellani et al., *Place and Health as Complex Systems*,
SpringerBriefs in Public Health, DOI 10.1007/978-3-319-09734-3_14

5. Our test of $C5$ shows just how nonlinear health outcomes are when trying to manage the negative impact that an environmental force like sprawl has on poor communities. Still, for most, the conclusions drawn from $C5$ are by no means new. What is new about the COP approach, however, is its attempt to make the issue of nonlinearity 'front and center' in our model building in order to construct more 'empirically accurate' health policy interventions.

6. Our test of $C6$ and $C8$ reminds scholars that power-relations, subjectivity, voice, and history are important dimensions of complex systems and need to be studied, whatever the topic. While such insights may seem minor to most social scientists, they remain outside the purview of the overwhelming majority of complexity scientists and computational modelers. As such, in terms of the Definitional Test of Complex Systems (DTCS), we see the inclusion of these two characteristics as both highly novel and significant.

7. Our test of $C9$ validated that studying agency adds an important tool for studying the role agency play in the dynamics of a complex systems issue like sprawl and health; plus, it hints at the potential of agent-based modeling for conducting thought experiments and theoretical validation. Still, the fictive nature of agent-based modeling, along with its struggle to model social structure, remain important caveats to address when considering the import of this approach.

In terms of the negative test of the DTCS, it seems that the COP definition, overall, fits well with our case study, both empirically and theoretically. There is, however, a key limitation to the COP definition: it does not conceptualize health or health care in complex systems terms. Health, as in the case of our current study, is still primarily treated as an outcome variable, rather than a system unto itself; neither is it seen as a factor that folds back to impact the compositional and contextual factors impacting it. In other words, health and compositional and contextual variables need to be developed into a more sophisticated k-dimensional profile (causal model). Furthermore, health care is not yet addressed as a complex sub-system within the larger system of place. Still, several of the articles in the 2013 special edition of *Social Science and Medicine*, such as Clark's "What Are the Components of Complex Interventions in Healthcare?," do begin to advance our understanding of this issue. Nonetheless, more needs to be done.

Finally, in terms of our attempt to advance Keshavarz et al [40], our study tried to make two important points. First, if the application of complexity science to the study of place is to more forward, scholars need to be more theoretically systematic in their usage of these ideas. The DTCS is an attempt to move scholars in this direction, offering a synthetically useful tool for testing the empirical validity and theoretical value of defining places as complex systems. The second point is that community health scholars desperately need to advance and integrate the latest developments in computational and complexity science methods. One such platform for making such an advance, which this study demonstrated, is the SACS Toolkit.

However, and by way of a conclusion, as with all research, additional exploration is necessary to determine if the COP approach, the DTCS and the SACS Toolkit are useful for modeling other public and community health topics.

References

1. Aldenderfer, M.S., Blashfield, R.K.: Cluster Analysis. Sage, Beverly Hills (1984)
2. Auchincloss, A., Diez-Roux, A.V.: A new tool for epidemiology: The usefulness of dynamic-agent models in understanding place effects on health. Am. J. Epidemiol. **168**, 1–8 (2008)
3. Castellani, B., Schimpf, C., Hafferty, F.: Medical Sociology and Case-Based Complexity Science: A Users Guide. Handbook of Systems and Complexity in Health, pp. 521–535. Springer, New York (2013)
4. Baçao Fernando, V.L., Painho, M.: Self-organizing maps assubstitutes for k-means clustering. ICCS 2005, LNCS **3516**, 476–483 (2005)
5. Bar-Yam, Y.: Dynamics of Complex Systems. Westview Press, New York (2003)
6. Bernard, P., Charafeddine, R., Frohlich, K.L., Daniel, M., Kestens, Y., Potvin, L.: Health inequalities and place: A theoretical conception of neighborhood. Soc. Sci. Med. **65**(9), 1839–1852 (2007)
7. Blackman, T.: Placing Health: Neighbourhood Renewal, Health Improvement and Complexity. The Policy Press, UK (2006)
8. Bowles, S., Durlaf, S., Hoff, K.: Poverty Traps. Princeton University Press, Princeton (2006)
9. Browning, C., Bjornstrom, E., Cagney, K.: Health and Mortality Consequences of the Physical Environment. Springer, Netherlands (2011)
10. Bruch, E., Mare, R.: Neighborhood choice and neighborhood change. Am. J. Soc **112**(3), 667–709 (2006)
11. Byrne, D.: Complexity Theory and the Social Sciences. Routledge, London (1998)
12. Byrne, D.: Complexity, configurations and cases. Theory Cult Soc. **22**(5), 95–111 (2005)
13. Byrne, D.: Complex realist and configurational approaches to cases: A radical synthesis.. In: Byrne, D., Ragin, C.C. (eds.) The Sage Handbook of Case-Based Methods. Sage, London (2009)
14. Byrne, D., Callaghan, G.: Complexity Theory and the Social Sciences: The State of the Art. Routledge, UK (2013)
15. Byrne, D., Ragin, C.: The Sage Handbook of Case-Based Methods. Sage, CA (2009)
16. Capra, F.: The Web of Life. Anchor Books Doubleday New York (1996)
17. Castellani, B., Hafferty, F.: Sociology and Complexity Science: A New Field of Inquiry. Springer, Germany (2009)
18. Castellani, B., Rajaram, R.: Case-based modeling and the SACS toolkit: A mathematical outline. Comput. Math. Organ. Theory **18**, 153–174(2012)
19. Cilliers, P.: Complexity and Postmodernism: Understanding Complex Systems. Routledge, New York (1998)
20. Cox, M., Boyle, P.J., Davey, P.G., Feng, Z., Morris, A.D.: Locality deprivation and type 2 diabetes incidence: A local test of relative inequalities. Soc. Sci. Med. **65**, 1953–1964 (2007)
21. Cummins, S., Curtis, S., Diez-Roux, A.V., Macintyre, S.: Understanding and representing place in health research: A relational approach. Soc. Sci. Med. **65**, 1825–1838 (2007)

22. Curtis, S., Setia, M. S., Quesnel-Vallee, A.: Socio-geographic mobility and health status: A longitudinal analysis using the national population health survey of canada. Soc. Sci. Med. **69**, 1845–1853 (2009)
23. Curtis, S., Rees-Jones, I.: Is there a place for geography in the analysis of health inequality? Soc. Health Illn. **20**, 645–672 (1998)
24. Curtis, S., Riva, M.: Health geographies i: Complexity theory and human health. Prog. Hum. Geogr. **a**, 1–9 (2009)
25. Curtis, S., Riva, M.: Health geographies ii: Complexity and healthcare systems and policy. Prog. Hum. Geogr. **b**, 1–8 (2009)
26. Du, K.L.: Clustering: A neural network approach. Neural Netw. **22**, 89–107 (2010)
27. Dunn, J., Cummins, S.: Placing health in context [editorial]. Soc. Sci. Med. **65**, 1821–1824 (2007)
28. Durie, R., Wyatt, K.: New communities, new relations: The impactof community organization on health outcomes. Soc. Sci. Med. **9**, 1928–1941 (2007)
29. Epstein, J.: Generative Social Science: Studies in Agent-Based Computational Modeling. Princeton University Press, Princeton (2007)
30. Gatrell, A.C.: Complexity theory and geographies of health: A critical assessment. Soc. Sci. Med. **60**, 2661–2671 (2005)
31. Gilbert, N., Troitzsch, K.: Simulation for Social Scientists, 2nd edn. Open University Press., Buckingham (2005)
32. Gladwell, M.: The Tipping Point : How Little Things Can Make a Big Difference. Little Brown, Boston (2000)
33. Gleick, J.: Chaos: Making a New Science. Penguin Books, New York (1987)
34. Haggis, T.: Approaching complexity: A commentary on keshavarz, nutbeam, rowling andkhavarpour. Soc. Sci. Med. **70**, 1475–1477 (2010)
35. Hammond, D.: The Science of Synthesis: Exploring the Social Implications of General Systems Theory. University Press of Colorado, Colorado (2003)
36. Holland, J.: Emergence: From Chaos to Order. Perseus Books, Cambridge, MA (1998)
37. Jain, A.: Data clustering: 50 years beyond k-means. Pattern Recognit. Lett. **31**(8), 651–666 2010
38. Kamada, T., Kawai, S.: An algorithm for drawing general undirected graphs. Inf. Process. Lett. **31**, 7–15 (1989)
39. Kauff man, S.A.: Investigations. Oxford University Press, Chicago (2000)
40. Keshavarz, N., Nutbeam, D., Rowling, L., Khavarpour, F.: Schools as social complex adaptive systems: A new way to understand the challenges of introducing the health promoting schools concept. Soc. Sci. Med. **70**, 1467–1474 (2010)
41. Klir, G.J.: Facets of Systems Science, 2nd edn. Kluwer Academic/Plenum Publishers, New York (2001)
42. Kluver, J., Kluver, C.: Social Understanding: On Hermeneutics, Geometrical Models and Artificial Intelligence, vol. 47. Springer (2011)
43. Kohonen, T.: Self-Organizing Maps, 3rd edn. Springer, New York (2001)
44. Kuo, R.J., Ho, L.M., Hu, C.M.: Integration of self-organizing feature map andk-means algorithm for market segmentation. Comput. Oper. Res. **29**, 1475–1493 (2002)
45. Lewin, R.: Complexity: Life at the Edge of Chaos. University of Chicago Press, Chicago (1992)
46. Lin-Fisher, B.: Forum fighting hospital plan: Akron general hosts even that argues against doctor-owned facility in northern summit. Akron Beac. J. **10-22-2008** (2008)
47. Macintyre, S., Ellaway, A., Cummins, S.: Place effects on health: How can we conceptualise, operationalise and measure them? Soc. Sci. Med. **55**, 125–139 (2002)
48. Maturana, H., Varela, F.: The Tree of Knowledge: The Biological Roots of Human Understanding. Shambala, Boston (1998)

49. Mitchell, M.: Complexity: A Guided Tour. Oxford University Press, New York (2009)
50. Morin, E.: On Complexity. Hampton Press, CressKill (2008)
51. Morrison, K.: Complexity theory, school leadership and management: Questions for theory and practice. Educ. Manag. Adm. Leadersh. **38**(3), 374–393 2010
52. Peterson, D., Brody, T., Davis, W., Goranson, S., Lafaire, M., Lee, J., Lehrman, L., Lutner, L., Zeller, A.: Urban sprawl modeling, air quality monitoring, and risk communication: The northeast ohio project. Environmental Protection Agency **EPA/625/R-02/016** (2002)
53. Pickett, K.E., Pearl, M.: Multilevel analysis of neighbourhood socioeconomiccontext and health outcomes: A critical review. J Epidemiol Community Health **55**, 111–122 (2001)
54. Rajaram, R., Vaidya, U., Fardad, M., Ganapathysubramanian, G.: Stability in the almost everywhere sense: A linear transfer operator approach. J Math. Anal. Appl. **368**, 144–156 (2010)
55. Rajaram, R., Castellani, B.: Modeling complex systems macroscopically: Case/agent-based modeling, synergetics and the continuity equation. Complexity **18**(2), 8–17 2012
56. Rajaram, R., Castellani, B.: The utility of nonequilibrium statistical mechanics, specifically transport theory, for modeling cohort data. Complexity (2014). doi: 10.1002/cplx.21512
57. Rajaram, R., Vaidya, U.: Robust stability analysis using lyapunov density. Int. J. Control **86**(6), 1077–1085 2013
58. Rajaram, R., Vaidya, U.G.: Lyapunov density for coupled systems. Appl. Anal. (2014). doi:10.1080/00036811.2014.886105
59. Rihoux, B., Ragin, C.: Configurational Comparative Methods: Qualitative Comparative Analsysi (QCA) and Related Techniques. Sage, Thousand Oaks (2009)
60. Riva, M., Curtis, S., Gauvin, L., Fagg, J.: Unravelling the extent of inequalities in health across urban and rural areas: Evidence from a national sample in england. Soc. Sci. Med. **68**, 654–663 (2009)
61. Robert, S.: Socioeconomic position and health: 'the independent contribution of community socioeconomic context'. Ann. Rev. Soc. **25**, 489–516 (1999)
62. Sampson, R., Morenoff, J., Gannon-Rowley, T.: Assessing 'neigborhoodeffects': Social processes and new directions in research. Ann. Rev. Soc. **28**, 443–478 (2002)
63. Smith, J., Jenks, C.: Qualitative Complexity. Routledge, UK (2006)
64. Smith, K.P., A., C.N.: Social networks and health. Ann. Rev. Soc. **34**, 405–429 (2008)
65. Sridharan, S., Tunstall, H., Lawder, R., Mitchell, R.: An exploratory spatial data analysis approach to understanding the relationship between deprivation and mortality in scotland. Soc. Sci. Med. **65**(9), 1942–1952 (2007)
66. Sturmberg, J., Martin, C.: Handbook of Systems and Complexity in Health. Springer, New York (2013)
67. Summit, H.: Healthy summit 2000 health indicators summary report. http://www.healthy summit.org
68. Uprichard, E.: Introducing Cluster Analysis: What It Can Teach us About the Case, pp. 132–147 Sage, CA (2009)
69. Urry, J.: Global Complexity. Blackwell Publishing. Oxford (2003)
70. Waldrop, M.M.: Complexity: The Emerging Science at the Edge of Order and Chaos. Simon and Schuster paperbacks, New York (1992)
71. Weaver, W.: Science and complexity. Am Sci. **36**, 536(1948)

Index

© Springer International Publishing Switzerland 2015

B. Castellani et al., *Place and Health as Complex Systems,*
SpringerBriefs in Public Health, DOI 10.1007/978-3-319-09734-3